COVID-19 & GLOBAL FOOD SECURITY

EDITED BY
Johan Swinnen
& John McDermott

International Food Policy Research Institute
1201 Eye Street, NW
Washington, DC 20005-3915 USA
www.ifpri.org

ISBN: 978-0-89629-387-8
DOI: https://doi.org/10.2499/p15738coll2.133762

Photo credits
COVER: Prabhat Kumar Verma/Shutterstock.
CHAPTER IMAGES: p. 6 Prabhat Kumar Verma/Shutterstock; p. 13 Atul Loke/Panos; p. 14 suprabhat/Shutterstock; p. 36 Atul Loke/Panos; p. 50 Mohsen Nabil/Shutterstock; p. 63 Mg Si Thu/Shutterstock; p. 64 Sundarrajan D/Shutterstock; p. 75 Sylvain Cherkaoui/Panos; p. 76 Andy Johnstone/Panos; p. 86 Atul Loke/Panos; p. 95 Stockpexel/Shutterstock; p. 96 Prabhat Kumar Verma/Shutterstock; p. 122 Jun Pinzon/Shutterstock.

Design and layout: Jason Chow
Project manager: Pamela Stedman-Edwards

Contents

Preface

Johan Swinnen and John McDermott

COVID-19 first emerged in China in late 2019, and the country mobilized a major epidemic response in January 2020, with stringent lockdowns and travel restrictions. The World Health Organization declared a global pandemic on March 11, and many countries soon began to impose measures to control the spread of the novel coronavirus. Since then, the disease has taken hundreds of thousands of lives and disrupted the livelihoods of billions of people.

The International Food Policy Research Institute (IFPRI) started a blog series on COVID-19 in February, first looking at the implications of the pandemic and responses in China. In March, as the enormous reach and potential impact of the pandemic became clearer, IFPRI researchers and guest authors began to look at its global repercussions on poverty and food and nutrition security. As the emergency response phase began in many low- and middle-income countries in the following months, we continued to use this blog series to analyze COVID-19 impacts. Entries in this series report results from continuously updated model assessments and innovative surveys – including phone surveys of households and firms and key informant networks – for early assessments of changes in food and nutrition security. At this point, there are over 40 entries in the COVID-19 series, and that number continues to grow.

IFPRI researchers have also developed a set of tracking tools that are publicly available. At the global level, these include trackers for staple foods, notably on price volatility and trade. In some African and South Asian countries, a daily food price monitoring system has been established in several local markets. To track and measure divergence in public policy responses, IFPRI has established a country-level COVID-19 policy response tracking system, a complement to the World Bank's social protection tracker.

This e-book compiles a selection of entries from the IFPRI blog series on COVID-19. The pieces provide key insights and analysis on how the global pandemic is affecting global poverty and food security and nutrition, food trade and supply chains, gender, employment, and a variety of policy interventions, as well as reflections on how we can use these lessons to better prepare for future pandemics. These pieces draw on a combination of conceptual arguments, global and country-level simulation models, in-country surveys, case studies, and expert opinions. Together, they present a comprehensive picture of the current and potential impact of COVID-19 and the policy responses to the pandemic on global food and nutrition security.

This book could not have come together without the editorial guidance of Pamela Stedman-Edwards and the design work of Jason Chow. We are grateful to Drew Sample and John McQuaid, who provided invaluable editorial support for the blog series from which this book is drawn, and support from all of IFPRI's Communications and Public Affairs Division, which made this work possible.

Finally, we dedicate this book to Rajul Pandya-Lorch upon her retirement. She has been a driving force behind many IFPRI publications and enthusiastically supported the creation of IFPRI's blog series on COVID-19 and the numerous associated virtual events. We will greatly miss her forward-looking leadership and strong commitment to IFPRI's work.

INTRODUCTION

1. COVID-19: Assessing impacts and policy responses for food and nutrition security

Johan Swinnen and John McDermott

COVID-19 has severely disrupted our lives, jeopardized the well-being of billions of people, and raised the specter of a global food crisis, all in just a few months. The huge impact expected on the world's economy and on global food security has been described in dramatic terms. The World Bank forecasts that the global economy will shrink by more than 5%, which would be the deepest recession since the Second World War. IFPRI researchers estimate that, in the absence of strong interventions in developing countries, the number of people in extreme poverty could increase by up to 150 million. The World Food Programme's executive director, David Beasley, has warned that the world is "not only facing a global health pandemic but also a global humanitarian catastrophe," and that, without action, COVID-19 could lead to "multiple famines of biblical proportions." Lawrence Haddad, executive director of GAIN, lamented that the coming food and nutrition crisis is not only biblical but "on steroids, and across generations."

Why such dire predictions? COVID-19 may not be as deadly as historical plagues or, more recently, the 1918–19 Spanish flu or Ebola, but it is unpredictable and highly transmissible, including by people who are asymptomatic but infected. This makes it difficult to control. Today's interconnected world has allowed the virus to spread with remarkable speed. On January 9, 2020, China officially recognized the first death from COVID-19, and by early March more than 100 countries were reporting cases (Figure 1). The number of officially reported cases has continued to increase, topping 10 million at the end of June. Over this period, the virus epicenter has shifted from China to Europe and the United States, and now many poor countries and regions including Africa, Latin America, and South Asia are facing rapidly rising infection rates and deaths.

More importantly, COVID-19 is a health crisis with multiple and widespread impacts on food systems, social systems, and economic development. The need to change daily practices and routines, many essential to livelihoods, and the consequent disruption of connections at local, regional, and global levels make the COVID-19 shock different from economic and climate shocks. And compared with previous pandemics, the much greater interconnection of trade and markets today – and the more complex nature of food, health, and economic systems – is amplifying the potential of COVID-19 to aggravate poverty and disrupt food systems. As a result, the impacts on well-being will be large relative to disease mortality rates.

The chapters in this book look across the broad range of impacts of this unprecedented crisis, providing forecasts, evidence, analysis, and recommendations for more effective policy responses to support food security. They draw on a combination of conceptual arguments, global and country-level simulation models, in-country surveys, case studies, and expert opinions. Key insights from the different contributions are the following.

The virus infections and public health responses to control COVID-19 transmission have had severe economic consequences. In an effort to control the disease, governments have imposed lockdowns that have shuttered many businesses, restricted travel within countries, closed borders to human traffic and trade in some food products, and imposed social distancing requirements and curfews that disrupt economic activity and force businesses and schools to close. The result has been economic recession, with spikes in unemployment, and major disruption of food systems and supply chains, as labor, transport, and trade are impeded.

The combined impact of recession and disruption is especially detrimental for the poor, who have been acutely affected by lost or decreased incomes and remittances as the pandemic has led to a severe economic recession in many countries. Simulations predicted that the poorest groups in society would see their relative incomes fall more than wealthier social groups, and early survey results from Ethiopia confirm these predictions. Poor people, whose main asset is their physical labor, have also suffered the most from the effects of lockdowns. The case studies on China (interruption of rural-urban migration), Egypt (declining tourism), and Myanmar (interruption of international migration) confirm that the economic and poverty effects of the fall in remittances due to lockdowns and travel restrictions are huge. Disruption of some of the public programs they rely on has aggravated the differential impacts on the poor. At the same time, however, many governments have responded with the introduction of new safety net programs and expansion of existing programs that can offset some of the lost income.

Women are likely to suffer more adverse impacts from the crisis. Income shocks and lockdowns have changed household and community gender dynamics and increased the disadvantages faced by women. Because of the potential for gender bias in government policies responding to COVID-19, it is crucial that social safety net programs explicitly account for gender effects in order to mitigate negative consequences during the emergency response period.

The severe disruption of food systems – including restrictions on labor and interruption of transport, processing, retailing, and input distribution – threatens the food and nutrition security of the poor. COVID-19 has exposed fragilities in food systems, especially in labor-intensive systems, such as those for fresh fruits and vegetables. The breakdown of supply chains due to the virus infection itself and a variety of policy restrictions has caused consumer prices to increase and producer prices to fall at the same time, increasing food insecurity for both urban and rural poor. Many poor people also suffered as their employment in food supply chains – transporting, marketing, and selling food – came to a halt. A series of country simulations of the food and economic sector impacts of COVID-19 (initial results are available for China, Egypt, South Africa, Myanmar, and India) show that income declines in food services and processing have been particularly strong. Impacts on farming itself are somewhat less severe, as many small farmers rely on family labor. However, it will be important to track the extent to which constrained access to key farm inputs and distorted prices affects investments in next years' production.

Early concerns over disruptions of global supply chains focused on the introduction of several trade constraints, in particular export bans in rice and wheat markets, by major exporting countries. Such export bans and trade restrictions threatened to exacerbate global food supply problems. Several of the initially introduced export constraints have since been removed. However, many other COVID-related government restrictions are causing unnecessary problems, both for trade between

Total confirmed COVID-19 cases

Limited testing means that the number of confirmed cases is lower than the actual number of cases.

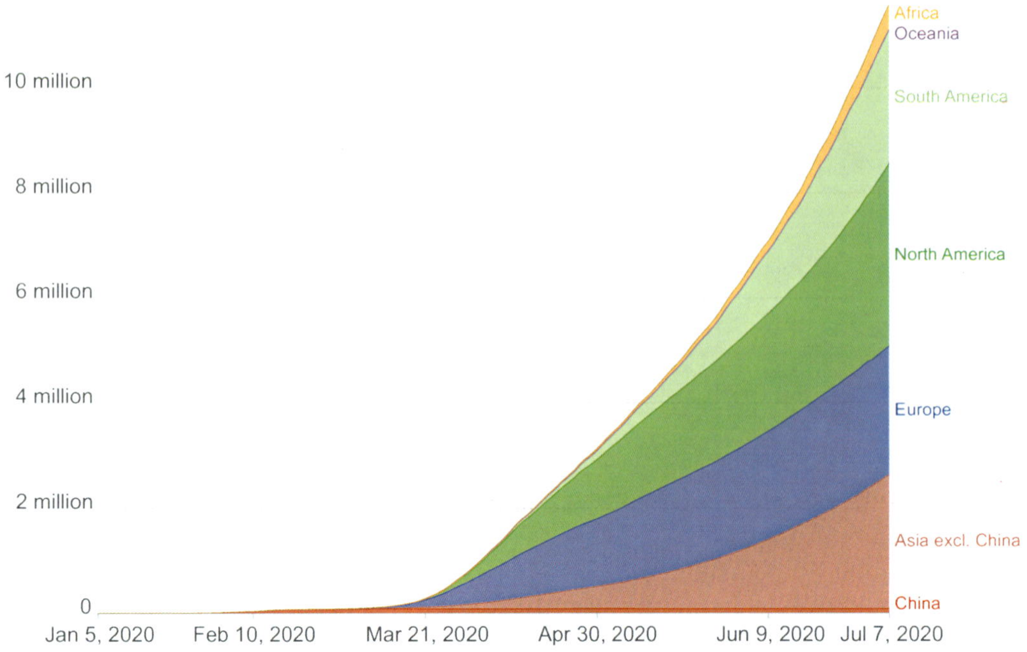

Source: European Centre for Disease Prevention and Control, Situation Update Worldwide, July 7, 2020.

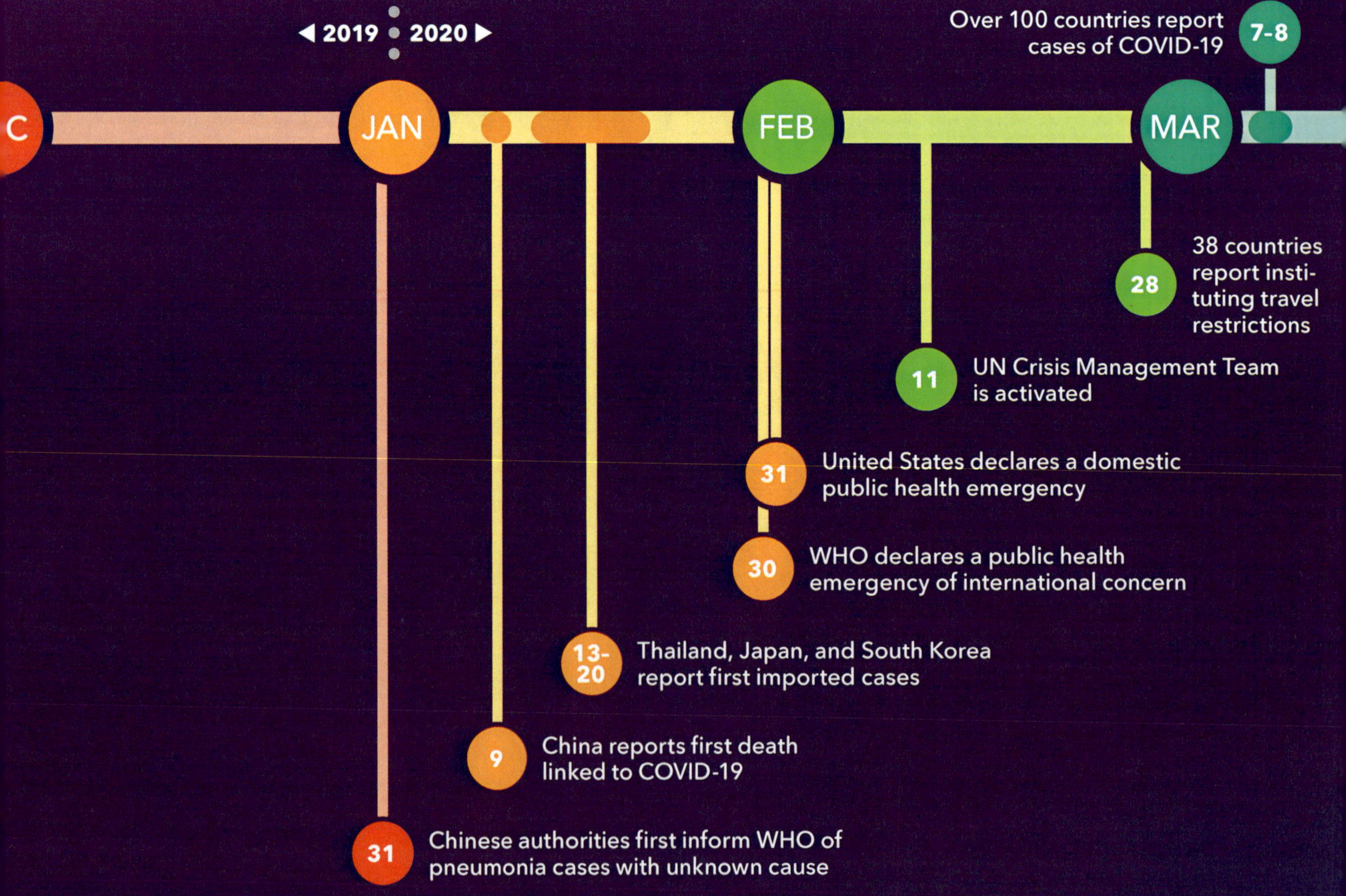

Timeline of global COVID-19 outbreak

25 India begins lockdown

APR

8 China lifts its lockdown on Wuhan

22 Outbreaks in Western Europe stabilize, upward trends continue in Africa, Central and South America, and Eastern Europe

MAY

13 UN Crisis Management Team warns against lifting lockdowns prematurely

21-22 COVID-19 cases surpass 100,000 on the African continent

JUN

22 Confirmed cases of COVID-19 surpass 9 million globally

J

Total confirmed COVID-19 deaths

Limited testing and challenges in the attribution of the cause of death mean that the number of confirmed deaths may not be an accurate count of the true number of deaths from COVID-19.

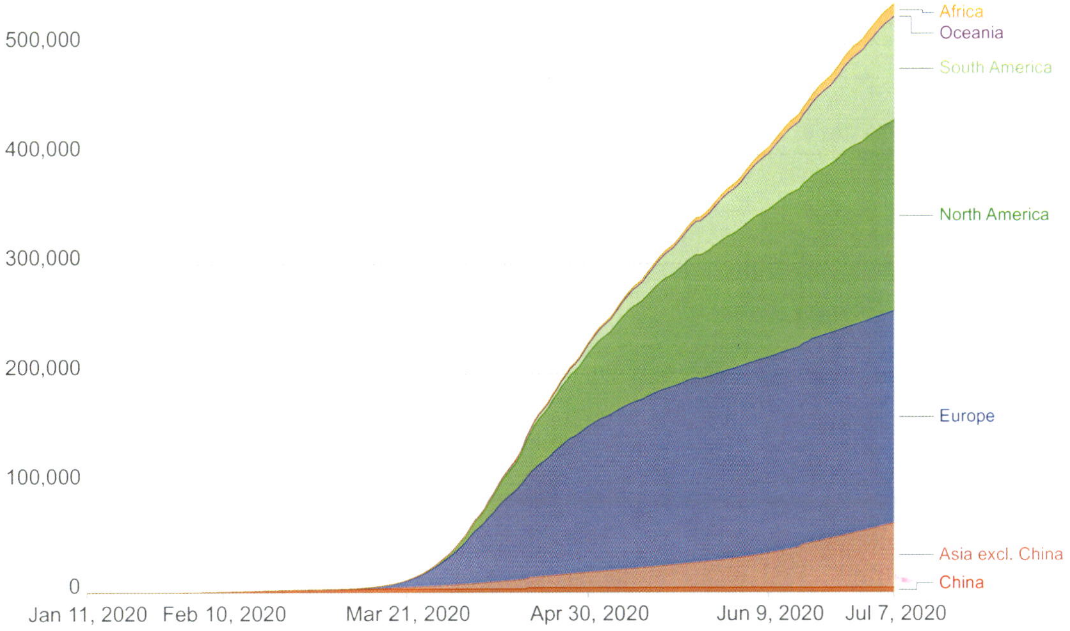

Source: European Centre for Disease Prevention and Control, Situation Update Worldwide, July 7, 2020.

developing countries and food marketing inside developing countries. In some countries, blanket policy actions, such as market and trade restrictions and curfews, are impeding food supply chains, while other countries were able to avoid these disruptions.

COVID-19 disruptions have also had negative consequences for long-term and hard-won progress on nutrition. Disruption of livelihoods and food supply chains means diets have become less healthy and nutrition programming for the poor and vulnerable has been interrupted. A simulation points to significant global declines in more nutritious but more expensive products such as fruits and vegetables and dairy products. Early survey results from Ethiopia on the vegetable and dairy supply chains confirm these simulation and expert predictions. Consumption of healthy diet ingredients has declined, and has declined most for the poorest groups in society. However, nutrition experts point to some viable options for mitigating negative nutrition consequences and for actions to hasten recovery and a return to the long-term positive trajectory of the past decade.

There are also other positive signs. While many food systems have been significantly disrupted, others have been more resilient, with food supplies relatively unaffected. Government responses to the crisis have varied widely, and some responses have had more positive effects than others. Innovation is occurring in many social programs and in NGO activities to overcome the constraints created by government regulations and health programs. Likewise, innovations and entrepreneurship in private food supply chains are helping to overcome obstacles and make food supply chains more resilient for the future. One example is the use of information and communications technology and e-commerce, which is growing at a much faster pace than expected prior to the COVID-19 outbreak. The country studies also predict significant economic rebounds in several countries once the lockdown measures are removed.

Looking forward, an intriguing question is about the next pandemics. In the recent past, epidemic/pandemic diseases have mainly emerged from Asia, but in the future disease emergence is likely to become more common and more disruptive in Africa as population densities, natural resource exploitation, and economic and social integration increase dramatically. The lessons we learn from COVID-19 should help us to design better policies and to build more resilient and inclusive food systems that limit the impact of future pandemics.

This collection of short essays provides a fairly comprehensive overview of where we are and what we know about how COVID-19 is affecting food security and livelihoods at the six-month mark. More specifically, they provide a detailed look at policy responses across the globe and how effectively they are working, along with recommendations for further innovations. In the coming months, new survey data, model simulations for more countries, and better insights on the spread of the virus, policy impacts, and innovations in public and private components of food systems will no doubt enhance our understanding. The IFPRI COVID-19 blog series will continue to report on new insights and on emerging issues, such as country policy responses and balancing their effects on health and economies; investments in short- and long-term responses to COVID-19 and their effects; reshaping food trade and supply chains in the recovery phase to manage continuing COVID risks; and evolving impacts on poor and vulnerable populations and how these can be monitored and mitigated.

SH MASALA
Delight

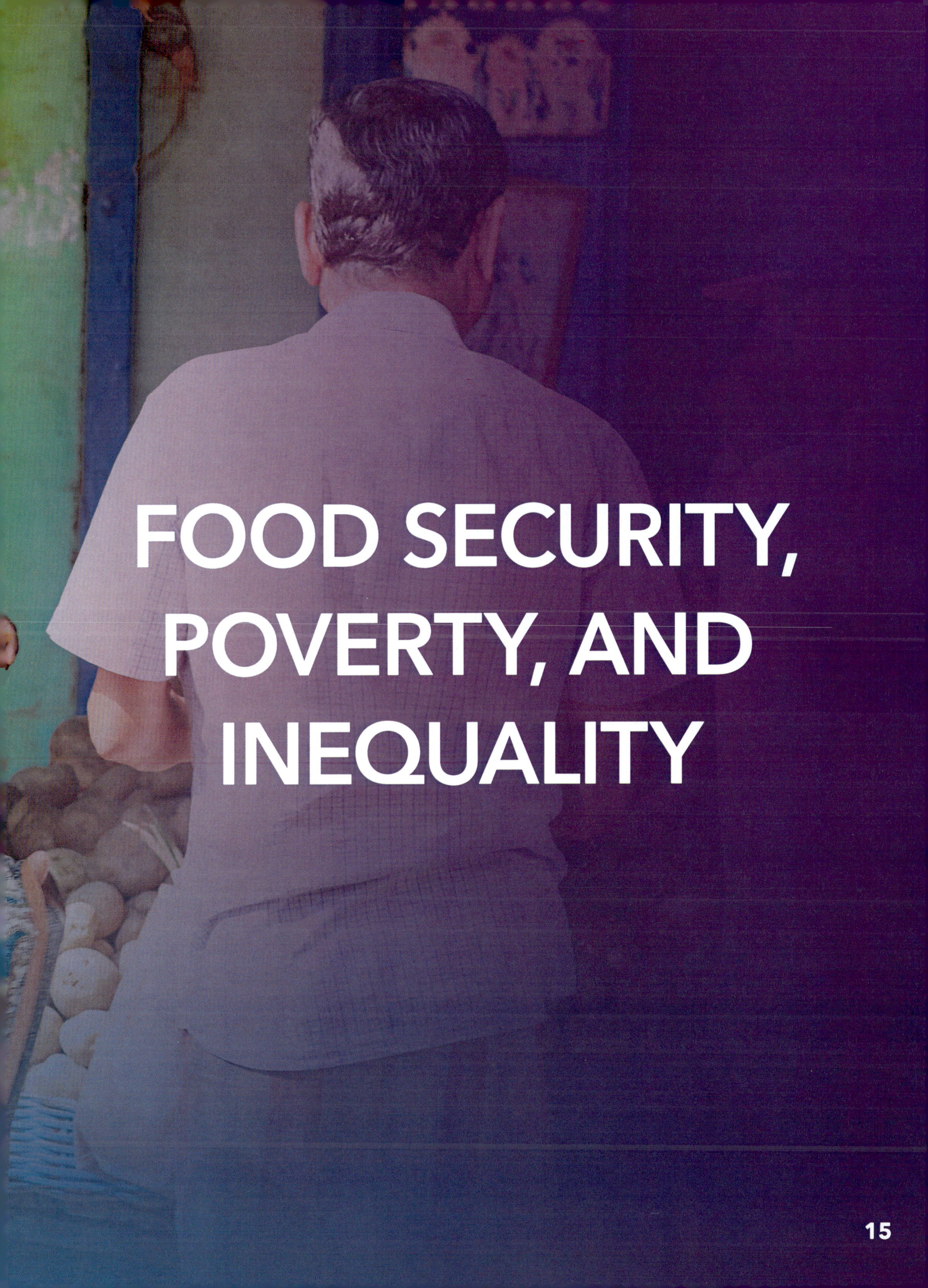

FOOD SECURITY, POVERTY, AND INEQUALITY

2. Poverty and food insecurity could grow dramatically as COVID-19 spreads

David Laborde, Will Martin, and Rob Vos

Cases of COVID-19 worldwide are growing exponentially, with major impacts on global staple food markets and poverty and hunger. On March 10, the number infections had just passed 110,000 with about 4,000 deaths. In the following month, the number of people with COVID-19 increased 20-fold (to over 2 million) and the number of deaths more than 30-fold (to over 135,000). The epicenter of the pandemic shifted from China to Europe and then to the United States. The coronavirus is now spreading rapidly in low- and middle-income countries, many of which lack robust health systems or strong social safety nets that can soften the pandemic's public health and economic impacts.

More than half of the world population is currently under some form of social distancing to contain the health crisis. As a result, millions of businesses have had to close shop. The International Labour Organization anticipates 200 million workers could be thrown into unemployment. In the United States alone, virtually overnight, 22 million people lost their jobs in early April. Governments in Europe and the United States have promised unprecedented fiscal and monetary stimulus measures to compensate for the income losses of businesses and workers and to contain an inevitable economic crisis. But the relief responses of low- and middle-income countries have thus far been more limited.

With COVID-19 and its economic fallout now spreading in the poorest parts of the world, many more people will become poor and food-insecure. In a new scenario analysis, we estimate that globally, absent interventions, over 140 million people could fall into extreme poverty (measured against the $1.90 poverty line) in 2020 – an increase of 20% from present levels. This in turn would drive up food insecurity. A global health crisis could thus cause a major food crisis – unless steps are taken to provide unprecedented economic emergency relief.

While considerable uncertainty surrounds the outcome, the world is very likely to face a deep recession in 2020 – at least as severe as the one following the global financial crisis of 2008-2009. A steep global economic downturn has already set in. Even assuming a strong rebound in the second half of the year on the back of unprecedented economic stimulus measures in the United States and Europe, the economic damage in major developed countries for 2020 is likely to exceed that of the Great Recession.

Using IFPRI's global model, we examined some of the likely impacts of the downturn for poverty worldwide and regionally.

Under the assumptions indicated in the box below, we project a downturn in global economic growth of 5% in 2020. This projection is broadly similar to the recent IMF forecast, which shows a downturn of the world economy from the 2% to 3% growth anticipated pre-pandemic to an actual decline of 3%.

TABLE 1 COVID-19 global economic recession in 2020

	PERCENTAGE CHANGE FROM BASE YEAR VALUES				
	Real GDP	**Household consumption**	**Export of goods** (value in constant dollars)	**Agrifood real value added**	**Agrifood exports** (value in constant dollars)
World	−5.0	−1.0	−20.9	−1.8	−24.8
Developed countries	−6.2	−0.1	−23.5	−3.1	−23.8
Developing countries	−3.6	−2.5	−18.0	+0.1	−30.5
Africa south of Sahara	−8.9	−3.2	−35.2	+3.9	−20.6
South Asia	−5.0	−3.7	−27.1	−2.0	−30.7
Southeast Asia	−7.0	−4.2	−27.7	−2.8	−31.9
Latin America	−5.9	−4.4	−30.8	−3.9	−28.5

Source: Authors, based on simulations with MIRAGRODEP model, April 2020 IFPRI global reference scenario.

Our scenario, however, indicates that the poorest nations face significantly greater adversity. The recession that has already started in Europe and the United States is projected to depress economic activity across developed countries by 6% on average in 2020, despite an expected rebound later in the year as social distancing measures are lifted and stimulus measures take effect. This recession will spill over to the rest of the world through lower demand for trade and lower commodity prices. Developing economies will be hurt by the economic fallout caused by their own social distancing measures and by increased morbidity affecting the labor supply for farming and other business activity.

For developing countries as a group, the economic fallout would lead to a decline of their aggregate GDP of 3.6%, but economies in Africa south of the Sahara, Southeast Asia, and Latin America would be hit much harder due to their relatively high dependence on trade and primary commodity exports. The recession is expected to be less severe in China and the rest of East Asia, where we expect the economic recovery to start sooner with the earlier lifting of containment measures.

We expect economies in Africa to be hit hardest (almost a 9% decline). But agrifood sectors may be spared and expand, as the collapse in export earnings and loss of capacity to import food push up domestic production. Lower labor demand in urban service sectors may push workers to return to agriculture, also contributing to greater domestic food production. With more workers in the sector, however, individual incomes would remain low.

FIGURE 1 Impact of COVID-19 global economic crisis on extreme poverty

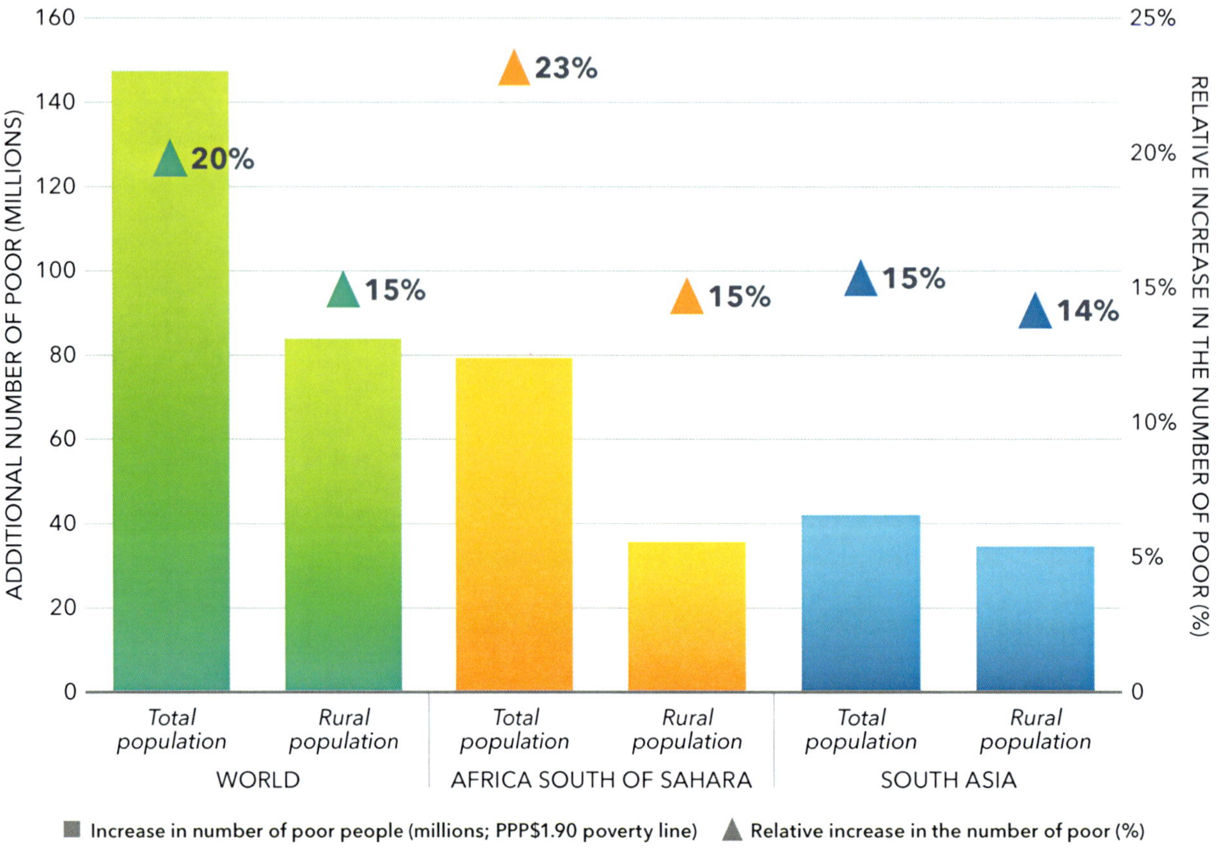

Source: Authors, based on simulations with MIRAGRODEP model.

Without social and economic mitigation measures such as fiscal stimulus and expansion of social safety nets, the impact on poverty would be devastating. (The scenario does not consider any such responses, as yet.) In addition to the 20% global increase in extreme poverty noted above, the scenario indicates urban and rural populations in Africa south of the Sahara would suffer most, as 80 million more people would join the ranks of the poor, a 23% increase. The number of poor in South Asia would increase by 15% or 42 million.

As these estimates refer to the extreme poor, that is, those who typically lack sufficient means to buy enough food, we expect a commensurate rise in the number of food-insecure people.

Heading off this dire outcome – a potential massive increase in global poverty and hunger – calls for an unprecedented policy response. High-income countries and international organizations should work to provide low- and middle-income countries with the necessary fiscal space and import capacity to expand health and social protection programs, strengthen food supply chains, and ensure adequate and affordable food supplies. Honoring the multiple calls for official and commercial debt relief would also help.

Key assumptions for the scenario analysis

- All countries affected by COVID-19 implement social distancing measures covering on average 40% to 50% of the population for between 2 and 3 months.
- International travel is essentially shut down, closing many tourism-related activities.
- Social distancing measures allow only essential work, such as food production and distribution, under normal conditions. We assume further that, on average, one-third of skilled workers can continue to work effectively via various forms of telecommuting.
- The containment measures cause bottlenecks and delays in international transport, pushing up freight costs by 3%.
- While the agriculture and food sectors have been identified as essential in most countries, we also assume some supply disruption caused by reduced labor mobility (for example, for seasonal migrant labor) and further, that perishable farm products suffer greater postharvest losses of 5% due to logistics problems and demand fallout.
- The present scenario accounts for the economic stimulus packages being implemented by countries in North America and in Europe, including significant income transfers to households. The scenario does not consider any additional international support or government stimulus in developing countries.

But even these measures are unlikely to be enough. There should also be concerted efforts to keep trade channels open to avoid piling an unnecessary food price crisis on top of the current health and economic disasters facing the world. We have developed a tracker to monitor these potentially damaging policy measures. In addition, large amounts of fresh, additional, low-conditionality funding will be needed to stave off a large-scale food crisis and invest in resilient food supply chains.

Detailed results of the scenario analysis can be found here and a detailed description of the underlying methodology can be found here.

This work was supported by the CGIAR Research Program on Policies, Institutions, and Markets (PIM), led by IFPRI, and United States Agency for International Development (USAID).

To read more about this topic, see: D. Laborde, W. Martin, J. Swinnen, and R. Vos. "COVID-19 Risks to Food Security." Science 369 (6503): 500-502.

Originally published April 16, 2020, and updated June 15, 2020.

3. COVID-19 is exacerbating inequalities in food security

Johan Swinnen

COVID-19 is disrupting economies and food systems everywhere, but the poor will suffer the greatest risk of food crisis. Based on model predictions, early empirical evidence, and lessons from previous crises, it is clear that the risk of increased food insecurity depends on the level of economic development. As employment and income opportunities fall for the poor, the gap between rich and poor is growing. Among the poor, urban poor and women are especially vulnerable.

The poor's food and nutrition security will be disproportionately impacted by COVID-19 because:

1. The global economic recession will have larger effects on poor people's incomes.

2. The poor spend a larger share of their income on food.

3. Among productive assets, physical labor – poor people's principal asset – will be most affected by COVID-19.

4. COVID-19 will cause more disruptions in private sector value chains in poor countries.

5. COVID-19 will cause disruptions in public sector programs for food, nutrition, health, and poverty, which are more important for poor people.

6. The poor have less access to sanitation facilities and healthcare.

7. Poor countries have lower economic capacities to compensate for declining incomes.

Economic models predict that under current conditions – relatively high food stocks, good harvests, low oil prices, and declining demand – global food prices are not going to rise. However, logistical problems in harvesting and transport will put upward pressure on food prices in some areas of the world. Paradoxically, the most important cause of rising food prices may be hoarding behavior by consumers and governments rather than market conditions. Despite several expert reports and economic advice not to repeat the same errors made during the 2007-2008 food crisis, many governments early on introduced trade constraints for foods (see IFPRI's Food Export Restrictions Tracker). Yet even absent a major rise in food prices, the food security situation of poor people is likely to decline significantly around the world.

Falling incomes and food and nutrition security

The global recession, caused by lockdowns and other restrictions on business activity to control COVID-19, leads to reductions in food consumption and declines in nutrition status – especially among the poor. Whether through rising food prices, falling incomes, or both, people have less real income to pay for their food and will adjust accordingly. This effect plays out more strongly the less income one has, meaning more hardship for the poor. Global models predict that for every percentage point of global economic slowdown, the number of people living in poverty would increase by 2% to 3%, or about 14 to 23 million worldwide. However, health and economic impacts may be much more pronounced in developing, rather than developed, countries with associated greater implications for poverty and hunger.

An early study by Rozelle et al. (2020) confirms these effects for rural households in China. Separated from their income sources as COVID-19 travel restrictions prevented them from working in urban wage jobs, workers suffered massive income losses totaling more than $100 billion. These families cut back significantly on nutrition. The majority of villagers are reducing spending on food, buying more grains and staples in bulk at low cost instead of more expensive goods like meat and produce.

Other early evidence confirms the disproportionate impact on the poor. Hirvonen et al. (2020) find that significant declines in incomes were greatest among the poor in Ethiopia; and Tesfaye et al. (2020) find that the decline in nutritious food in Ethiopia was also more important among the poor.

COVID-19 impacts the poorest particularly hard because it directly affects their most important, sometimes only, productive asset: labor, especially physical labor. Richer people typically have a portfolio of productive assets, such as capital and land, and their labor is typically of a different quality. Even while locked down inside a townhouse or a city apartment, they can work via computer over the internet, spending their productive hours on email and Zoom. This is not the case for poor people with low skills whose only source of income is likely to entail leaving home to do manual work.

Within the group of "the poor," women and urban consumers particularly seem to suffer more. Quisumbing et al. identify several reasons for the gendered impact of COVID-19. A series of recent country studies by IFPRI researchers, summarized in Thurlow (2020), shows how urban poor who have lost their jobs and face the same or higher food prices are particularly negatively affected.

Another group of poor people that are among the hardest-hit by COVID-19 restrictions are those who have to travel for work. Studies show large negative effects for migrant workers in several categories: rural-urban (as in China–see Rozelle et al. 2020), international (as in Egypt–see Breisinger et al. 2020; and Myanmar–see Diao and Wang 2020), or rural-rural (as in India, where landless workers travel to work in seasonal jobs such as harvesting–see Dev 2020).

Disruption of private value chains

Harvesting may be disrupted because of a lack of workers; planting because of a lack of seed or fertilizer; transport because of reduced transport facilities; and market exchange because of lockdowns or social distancing. What we are witnessing is a disruption of the food system similar to what happened during 1990s-era transition processes when supply chains collapsed. Those experiences showed that impacts were strongly heterogeneous, depending on the nature of the commodity, the resource-intensity of the systems, and the level of economic development.

But in the 1990s, the key production factor affected was capital (see Rozelle and Swinnen 2004). Today, as noted, the disruption is mostly related to labor constraints. As a consequence, capital-intensive food value chains (mostly in rich countries, or in richer parts of poor countries) are much less affected than labor-intensive value chains (mostly in poor countries). Reardon et al. (2020) point out that the impacts will be greatest in informal-sector small and medium-sized enterprises, which are labor-intensive with high densities of workers in small spaces. Modern retail and food-service firms face fewer problems. Again, these differences are affecting food security among the poor disproportionately.

Disruptions in public sector programs and less access to health and sanitation facilities

COVID-19 will cause disruptions in public sector programs for food, nutrition, health, and poverty, which poor people depend on. For instance, India's national lockdown regulations implied closing of schools. This means that school feeding programs – one of the country's largest safety nets – have been suspended. Other safety nets are also affected, including nutrition programs in community courtyard sessions for pregnant women and lactating mothers. Key health programs, such as child immunization, have been disrupted as well. And of course, public food relief programs face the risk of exposing more people to the virus by attracting large crowds at distribution points.

These disruptions affect the poor disproportionately. They compound the problems associated with unequal access to sanitation facilities, including basic facilities such as running water, which is crucial for protection against COVID-19, and limited access to healthcare. Good healthcare institutions are less available in poor countries in general and access to them is particularly limited for the poor.

Limited government capacity to compensate

While many developed countries have responded to the economic fallout from COVID-19 by ramping up spending and using monetary policies, options for developing countries may be more limited. Developing countries will need to prioritize, focusing their responses on health, essential goods and services, the domestic financial circuit in local currency, and the foreign currency market linked to international trade and external debt. Such a focused approach can help finance public spending on programs like cash transfers and safety nets for the poor and vulnerable, and public investments to keep firms operating. At the same time, the international community also has a crucial role to play in supporting countries in their policy responses, including through international organizations like the World Bank, United Nations, and IMF, as well as the multilateral development banks.

In summary, several compounding factors mean that COVID-19 is likely to cause another major food crisis among the poor. To avoid a food crisis, governments will need to implement policies and programs that target those most impacted and help address the negative impacts.

Originally published April 10, 2020, and updated June 20, 2020.. This post also appears on the Global Alliance for Improved Nutrition (GAIN) Nutrition Connect blog.

4. COVID-19 lockdowns have imposed substantial economic costs on countries in Africa

James Thurlow

It is too soon to assess the full economic impacts that COVID-19 lockdowns will have on developing countries. But early research indicates that many African economies are significantly impacted and that poorer households are struggling.

IFPRI is conducting a series of country studies, in collaboration with local and government partners, that use economywide models to estimate the impacts of lockdowns, assess the exposure of food systems, and identify vulnerable population groups.

This research is ongoing and the situation is evolving rapidly, but three clear findings are emerging:

- Developing countries are shouldering substantial economic costs.
- Food supply chains are exposed, despite being largely exempted from lockdowns.
- Nonpoor urban households face the largest income losses, but poverty is rising sharply.

Developing countries are shouldering substantial economic costs

We do not yet know how long lockdowns will remain in effect, what their full impact will be in 2020, or how quickly African economies will recover from these shocks.

But findings from our country studies show that the current crisis is leading to much larger and more rapid contractions of economic activity than seen in previous crises, including the global food crisis of 2007–2008 and the 2009 recession. In addition, unlike in previous crises, it is domestic policies, rather than global shocks, such as trade disruptions and reduced remittances, that are imposing most of the economic costs, at least for now.

Our findings are alarming: In Nigeria, Africa's largest economy, we estimate a 38% drop in GDP during the five-week lockdown from late March to the end of April. South Africa appears to be experiencing a similar-sized shock. Impacts are also large in Ghana, where GDP fell by 28% during that country's three-week lockdown.

The enormous economic costs of the lockdowns are making it difficult for governments to maintain support for these policies. But any easing of restrictions must balance the economic importance of various sectors with the risks posed to those who work in them. IFPRI is working with governments to understand these trade-offs and avoid disruptions to national food systems.

Food supply chains are exposed, despite being largely exempted from lockdowns

Most African governments consider food supply chains to be "essential" and have exempted them from lockdown policies. However, while food may be exempt, food systems are not immune to the effects of the pandemic.

In Nigeria, for example, we estimate an 18% decline in agrifood GDP during its five-week lockdown, and a 20% decline in Ghana during its three-week lockdown. While other sectors, such as manufacturing and construction, are suffering even larger declines, the food supply chain is particularly important for poor workers and consumers.

Some impacts on the food system are direct. The closure of hotels, restaurants, and bars was an early and common target for most lockdown policies in Africa. That said, while eating meals prepared outside the home is important for many urban consumers, it comprises a small part of the overall food economy.

Most impacts on food systems are indirect, and mainly caused by falling incomes. Even when farmers, food processors, and traders are exempt from lockdowns, they may still be unable to sell products if consumer incomes and demand for food decline.

So far, food supply chains are faring better than other parts of the economy in most countries. But this could change. In Nigeria, the government has closed some food markets in Lagos and restricted food trading times in major cities to only four hours every other day. If this prevents consumers from getting access to food, it could quickly overshadow other disruptions to the food system and become a major source of economic costs across the entire economy.

For now, this remains a crisis of food access driven by income losses, rather than one of food availability. But food supplies could become more of a concern if lockdowns go on for longer, if they are applied to rural areas, or if the movement of rural workers is restricted.

Nonpoor urban households face the largest income losses, but poverty is rising sharply

It may seem counterintuitive that Africa's nonpoor urban households are hardest hit. But manufacturing and business services are facing the strictest lockdowns in most places, and these sectors are often concentrated in cities and employ better-educated workers.

In Nigeria, for example, incomes in the top quintile are estimated to fall by 41% during the lockdown, while the bottom quintile's incomes fall by 23%. There are similarly uneven distributional impacts in Ghana and South Africa.

But higher-income households are better able to offset such losses by drawing on savings and other assets. Poor and rural households are also suffering substantial losses, and for them, even a small drop in income can have detrimental and lasting effects.

More concerning is that the number of poor people is rising. In Nigeria and Ghana, for example, we estimate that national poverty rates will increase by 15 and 13 percentage points, respectively – that means 30 million more Nigerians and 4 million more Ghanaians living on less than $1.90 per day.

As the situation evolves, it could bring even greater losses to Africa's poor. Tighter restrictions on urban markets, for example, could shift more of the burden onto poor consumers and smallholder farmers.

Reassessing priorities, while maintaining a focus on the poor

The economic losses brought by the current crisis are much deeper and occurring more quickly than those brought on by earlier crises. Governments are under enormous pressure to provide short-term emergency relief while also planning and investing in economic recovery.

Financing emergency and recovery programs will be hampered by lower tax revenues caused by lockdown policies, and by countries' limited ability to borrow due to uncertainty in global markets and large debts accumulated since the last crisis. Some displacement of pre-COVID-19 policies and priorities is inevitable.

As governments reassess their policy priorities, they should not lose focus on the longer-term growth and poverty reduction that have underpinned Africa's past decade of strong economic development. Minimizing both the economic and health impacts of the COVID-19 pandemic will require coordinating both health and food system policies, and ensuring that they work for the poor. The outbreak of COVID-19 has confirmed the importance of having well-functioning health and food systems – achieving this requires sustained investment, even during times of crisis.

The work discussed in this blog post was supported by the CGIAR Research Program on Policies, Institutions, and Markets led by IFPRI. Originally published May 8, 2020.

5. How China can address threats to food and nutrition security from the COVID-19 outbreak

Kevin Chen, Yumei Zhang, Yue Zhan, Shenggen Fan, and Wei Si

Since the outbreak of the COVID-19 pandemic, China's national and local governments have adopted stringent mitigation policies, including mandatory lockdowns, suspension of public transportation, and travel restrictions. While these measures are necessary, they could potentially lead to hiccups in food and nutrition security. Pandemics like Ebola, Severe Acute Respiratory Syndrome (SARS), and Middle East Respiratory Syndrome (MERS) all had negative impacts on food and nutrition security – particularly for vulnerable populations including children, women, the elderly, and the poor. For example, when Ebola first hit Guinea, Liberia, and Sierra Leone in 2014, rice prices in those countries increased by more than 30%; the price of cassava, a major staple in Liberia, skyrocketed by 150%. In 2003, the SARS outbreak delayed China's winter wheat harvest by two weeks, triggering food market panics in Guangdong and Zhejiang, though production and prices were largely unaffected in the rest of China.

Since the beginning of the outbreak in late December, food prices have remained stable in Wuhan, in Hubei province – and in fact, all over China. Supplies of staples, fruits, vegetables, and meats have been adequate despite sporadic reports of price hikes and shortages in isolated locations. But there is no room for complacency. Media reports indicate that the poultry industry is already under stress due to a lack of adequate feed supply and interruptions in the timely marketing of its products. If nothing is done, the poultry supply could begin tightening, and these problems could spread to other industries – creating a food supply hiccup and a threat to food and nutrition security for many.

The current situation

Disruptions are not expected to be severe, as the food supply has been sufficient and the market has been basically stable, at least so far. Wuhan has been under lockdown since January 23 to contain the spread of the virus. Restrictions on transportation and movement of people have led to some food logistics challenges. However, as the lockdown began just before the Chinese New Year on January 25, most citizens had already stocked food, and businesses had established reserves of goods for the holidays, so food prices have not been significantly affected. In February, China's consumer price index, a gauge of inflation, went up 5.2% year-on-year. But food prices surged 21.9%, largely due to pork price hikes triggered by the impact of African swine fever on hog production. On a month-on-month basis, national consumer prices rose 0.8%, while food prices increased 4.3%, led by an uptick in the prices of fresh vegetables and meat. However, in March, both the consumer price index and food prices index declined on a month-on-month basis, falling 1.2% and 2.7%, respectively. The price of fresh vegetables has returned to normal after a surge in February, indicating a modest impact of COVID-19 on China's food market.

The poultry industry has suffered more adverse effects. Transportation blockages create difficulties for distribution of inputs like feed, and some firms have already encountered input shortages, difficulties in product delivery, and labor shortages. The ban on the movement of live poultry (believed to be a potential disease risk) has stopped farmers from getting chickens and eggs to market, and has led some to bury chicks and ducklings alive. According to industry estimates, market input of chickens and ducklings has decreased by about 50%. Coupled with the extended impact of African swine fever, supplies of meat could plunge. According to data from the National Bureau of Statistics, pork production dropped by 29.1% and total meat production decreased by 19.5% in the first quarter year-on-year.

The food system beyond agriculture has also been significantly affected, and these impacts will grow if processing enterprises cannot restart production in the near future. Only 24% of agricultural products are directly consumed by households, while 77% are used as intermediate inputs, 41% go to food processing enterprises, and 3% are used by restaurants. In the wake of the coronavirus outbreak, many orders were canceled and many restaurants had to close their doors. The supply of processed foods remained relatively abundant. According to data from the National Bureau, production in food industries – such as sugar, meat, and rice processing – decreased only modestly or maintained growth in February. But production may also be affected by a lack of workers and falling demand for agricultural products. One priority, then, has been allowing migrant workers to return to these jobs.

Production of staple food crops such as wheat, rice, and vegetables is expected to remain stable. The 2014 Ebola epidemic led to an increase in abandoned agricultural areas and reduced fertilizer use in West Africa. If staple food production is affected, the impact on food security could be grave. China is seeing reassuring signs for staple foods, with spring planting going smoothly, and recorded a 3.5% year-on-year increase in the added value of the country's planting industry and had sufficient daily food supplies.

Domestic and international trade disruptions may trigger food market panics. During the 2003 SARS outbreak, panic-buying of food and other essentials hit many places in China. If this happens again, it would exacerbate temporary food shortages, lead to price spikes, and disrupt markets. If not controlled quickly, food panics can spread and threaten broader social stability. Export restrictions and the potential imposition of nontariff trade barriers, on the premise of safety concerns, would also exert negative impact on food supply chains. The 2007–2008 food price crisis reminds us that export bans can drive food prices up and cause volatility. Unfortunately, 11 countries currently have active binding export restrictions on food.

Restrictions on mobility may lead to labor shortages. Many companies have given workers extended leave in response to the outbreak; this could leave many manufacturing and service enterprises without enough workers. Large numbers of migrant workers who returned to their hometowns for the New Year break were trapped there by quarantine measures. The number of migrant workers who returned to work declined by 30% at the end of February in 2020 compared to the same period in 2019. The resulting labor shortage will likely impact both domestic and global supply chains.

How has China responded?

Fortunately, the government has been targeting these problems since the early stage of the outbreak. China's earlier responses to ensure food security are discussed in a forthcoming article for *China Agricultural Economic Review*'s special section on "Agriculture and Food Security under Novel Coronavirus Pneumonia (NCP) Emergency" (Chen, Fan, and Zhan 2020).

On February 5, Premier Li Keqiang called on ministries to coordinate to ensure an ample supply of food and effective logistics for delivering agricultural inputs, emphasizing the responsibility of local governors. To ensure the smooth logistical operation of regional agricultural and food supply chains, China has opened a "green channel" for fresh agricultural products and prohibited unauthorized roadblocks. The Ministry of Agriculture and Rural Affairs (MARA), Ministry of Transport, and Ministry of Public Security jointly issued a notice on January 30 urging relevant departments to coordinate to ensure effective logistics for agricultural products and materials. MARA issued a further emergency notice on February 4, calling on these departments to maintain market functions and ample supplies of meat, eggs, and milk.

To address the challenges facing the livestock sector, the National Development and Reform Commission and MARA jointly issued the "Notice on Promoting Multiple Measures and Promoting the Expansion of Production Guaranteed Supply of Poultry and Aquatic Products" to accelerate the resumption of production. The notice recommends gradual reopening of live poultry markets. The government also supports the construction of centralized slaughtering points, cold chain logistics, and other infrastructure to improve value chains. Feed production and slaughter enterprises are required to accelerate production in order to restore and increase the effective supply of livestock and poultry products. They are also being provided with production guidance and technical services to strengthen animal and plant epidemic prevention and control. The government also introduced financial supports for food production to prevent a decrease in the credit balances of agriculture-related enterprises and reduce their financing costs. The Agricultural Bank of China has strengthened its services to support 349 key enterprises to ensure stable production and supply of agricultural products, with a loan balance of 41.4 billion yuan.

The burden on farming enterprises is mitigated by reducing or deferring their tax payments, reducing their rent, and deferring payment of their social insurance premiums. For example, the China Banking and Insurance Regulatory Commission issued the "Circular on Implementing Provisional Postponement in Principal and Interest Repayment for Loans to SMEs and Micro Enterprises (No.6)." Epidemic-hit small and medium-sized enterprises (SMEs) and micro firms, including small business owners and individual household businesses, can apply to their banks to defer repayment of principal and interest payable from January 25 to June 30, 2020; SMEs in China's Social Security Schemes are exempt from making employer contributions to pension, unemployment, and work-related injury insurance schemes between February and June 2020.

The use of e-commerce and delivery companies is another important means to ensure the food supply. As lockdown measures led to a huge spike in demand for home delivery of fresh groceries, e-commerce companies in China announced an in-app feature for contactless delivery, allowing a courier to leave an order in a convenient spot for customer pick-up, without person-to-person interaction. The use of these delivery platforms has helped resolve logistical challenges, while minimizing the potential risk of infection from visiting crowded food markets.

Finally, ensuring food security requires a means to address the loss of workers' incomes caused by the interruption of economic activities, as these income effects may lead to drastic reductions in nutrition, especially for vulnerable groups. MARA and the Ministry of Human Resources and Social Security issued the "Circular on Implementation Plan" on March 26 for expanding local employment of returning rural migrant workers. The plan aims to promote local employment of migrant workers in agricultural production, as well as to help migrant workers return to work while ensuring their safe movement. A number of public welfare jobs were set up for migrant workers who face difficultly finding jobs on the market.

Looking forward

It's unclear how long the outbreak will last globally. As lockdowns and social distancing measures are implemented in the rest of the world, evidence and lessons from China not only have important policy implications for China to ensure a robust food system, but can also provide insight for other countries to help prevent food and nutrition security crises.

Continue to closely monitor food prices and strengthen market supervision, particularly in Wuhan and across Hubei and nearby provinces. Transparent market information will enhance the government's overall management of the food market and help to prevent the onset of panics, and can guide farmers in making rational production decisions. Potential for speculation remains at every stage of the supply chain, so there is also a need for sound market supervision. For processed products, supervision is also important to maintain food quality and safety.

Ensure smooth logistical operations of regional agricultural and food supply chains. China opened a "green channel" for fresh agricultural products and banned unauthorized roadblocks to ensure normal function of food supply chains. In addition, e-commerce and delivery companies can also play key logistical roles. For example, as lockdown measures have led to a huge spike in demand for home delivery of fresh groceries, e-commerce companies have announced an in-app feature for contactless delivery, allowing the courier to leave an order in a convenient spot for contactless customer pick-up. Making use of these delivery platforms could address many logistical challenges for obtaining food, while minimizing the potential risk of infection from visiting crowded markets to buy groceries.

Ensure the smooth flow of trade and make full use of the international market as a vital tool to secure food supply and demand. After reaching the phase-one trade deal with the United States to buy $40 to $50 billion of US farm products annually for the next two years, China has further announced it is cutting tariffs on US goods to ensure supplies and to alleviate economic and trade frictions. Meanwhile, increasing livestock product imports could help to buttress supply and stabilize the prices of domestic livestock products.

Protect vulnerable groups and provide employment services to migrant workers. International experience shows that the impacts of pandemics fall disproportionately on vulnerable populations, including children, pregnant women, elderly people, malnourished people, and people who are ill or immunocompromised. If many workers are unable to earn an income due to the disruptions of the outbreak, that risks an increase in poverty. As the government manages the epidemic response, it will be essential to restore the income streams of migrant workers and normal business operations. Workers are encouraged to find jobs near their homes, and migration should be managed to prioritize their health. Policies to match workers with companies are being implemented. But to work, these policies will require close supervision.

Regulate wild food markets to curb the source of the disease. Many zoonotic diseases originate from wildlife; HIV, Ebola, MERS, and SARS have all made the leap from wildlife to humans, spawning international outbreaks. On January 26, the government announced a temporary ban on wildlife trade from markets, restaurants, and e-commerce until the epidemic is over. To avoid future epidemics, on February 24, the government further issued the decision to ban the trade and consumption of wildlife as food. Use of wild animals for medicine, pets, and scientific research will be subject to strict examination and approval by relevant departments.

Originally published February 12, 2020, and updated June 10, 2020.

For more on China's early responses to COVID-19, see Z. Chen, S. Fan, and Y. Zhan, "COVID-19 and Food Security: Early Responses, Impact, and Lessons from China," China Agricultural Economic Review, 2020 (forthcoming).

6. Assessing the toll of COVID-19 lockdown measures on the South African economy

Channing Arndt, Sherwin Gabriel, and Sherman Robinson

In trying to limit the spread of COVID-19, policymakers are confronting the difficult task of balancing the positive health effects of lockdowns against their economic costs – particularly the burdens imposed on low-income and food-insecure households.

South African lockdown policies are relatively stringent, and the economic impacts are large. Figure 1 presents impacts on the income components of gross domestic product (GDP), based on an analysis using a social accounting matrix (SAM) model, a tool well-suited to assessing the impacts of short-term shocks. The work is a collaboration between IFPRI, the National Treasury of South Africa, the South African Reserve Bank, and UNU-WIDER.

GDP can be viewed as a flow of goods and services. The lockdown has direct effects that restrict this flow. Prominently, there is a forced reduction in production, and final demand for goods and services falls as businesses and households are locked down. Indirect effects follow. For example, because many business operations, including some in manufacturing, are reduced to operating at low levels or not at all, demand for electricity declines, which in turn reduces demand for coal. Across productive sectors and households, these indirect effects propagate throughout the economy. The highly disaggregated SAM model assesses direct and indirect effects across these multiple sectors.

Once all indirect effects of the lockdown are considered, the total flow of goods and services is reduced by about a third (see righthand bar in Figure 1), with indirect effects accounting for most of the reduction. Figure 1 also shows how this reduction is distributed across wage earners (divided into categories by educational attainment) and returns to capital. These declines in earnings should be interpreted as being due to reductions in hours worked and in the rate of utilization of factories, machines, and other elements of installed capital. Note that the negative impacts on wage earnings are larger for less-educated workers.

In South Africa, then, COVID-19 public health responses have very large implications for economic activity and income, with especially strong implications for households with low education levels who depend on wage earnings.

On their own, these negative economic shocks are sufficiently large to push many households into food insecurity. To borrow a term from Amartya Sen, the lockdown could be characterized as a policy-induced reduction in household "capabilities." Increased food insecurity results principally from the severe shock to household incomes rather than a shock to food availability such as occurs in a drought.

Because the source of food insecurity is a collapse in earnings, income transfers via social protection are highly effective in countering the economic effects of lockdowns. In South Africa, government

FIGURE 1 Lockdown impacts on wage earnings and income GDP components in South Africa, as percentage deviation from pre-crisis levels

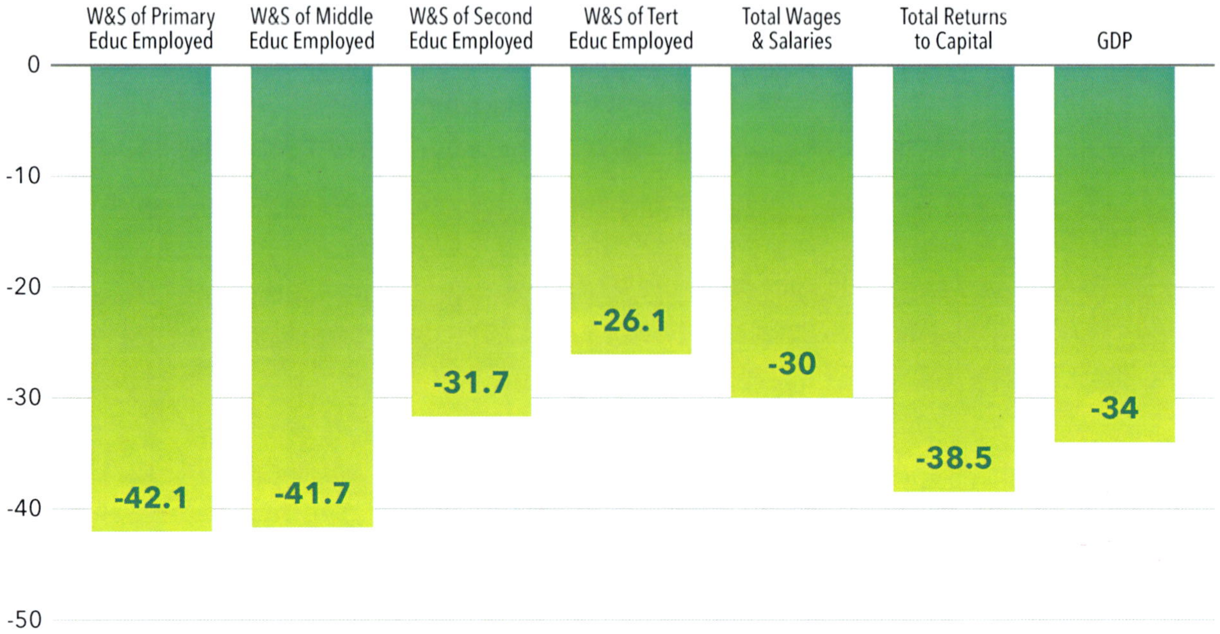

Source: Social accounting matrix (SAM) model.

Note: W&S = wages and salaries.

transfers are helping to substantially support total income of households in the lower half of the income distribution, blunting (but far from offsetting) the impacts of the crisis.

Using an income distribution and food security lens, the remarkably rapid and severe shocks imposed because of COVID-19 illustrate the value of having channels in place to transfer income to vulnerable households. This provides policymakers with the ability to soften the impacts of such "black swan" shocks on vulnerable populations. Looking ahead, preparing for future shocks also requires that, in good times, countries build their fiscal resources so that they can respond adequately at times of crisis.

Attention should now turn to developing a longer-run strategy for navigating the pandemic. Loosening movement restrictions too quickly risks a rapid increase in infections that could overwhelm the health system and cause more economic shocks as many workers become ill. South Africa's large number of people with HIV has resulted in a wealth of experience in infectious disease on which it can draw. HIV is also a potential risk factor for COVID-19 complications, presenting additional public health challenges. Overall, decisions on public finances will need to carefully consider how to address multiple spending demands with lower tax revenues. Devising suitable responses will require that economists and epidemiologists work together to understand the mechanisms at work and balance the health dimensions of policies to contain the pandemic and their economic fallout – especially for vulnerable groups.

This work took place within the framework of the Towards Inclusive Economic Development in Southern Africa (SA-TIED) program. Support to SA-TIED from the National Treasury of South Africa, the European Union, UNU-WIDER, and the CGIAR Research Program on Policies, Institutions, and Markets (PIM) is gratefully acknowledged.

Originally published May 6, 2020.

7. Addressing COVID-19 impacts on agriculture, food security, and livelihoods in India

S. Mahendra Dev

India took early action to limit the spread of COVID-19, ordering a 21-day nationwide lockdown for its population of 1.3 billion people starting March 25. Subsequently the lockdown was extended three more times before May 31. The unlocking of India began June 1, except in containment zones. The novel coronavirus has spread widely in India and the number of reported infections is 217,000, with relatively few deaths, at 6,075, as of June 4. However, as COVID-19 cases are increasing fast, there is great concern about the disease's potential spread and impact. India has to be ready for a possible surge. The government views the pattern of the spread of COVID-19 as similar to the 2009 H1N1 influenza pandemic, meaning the spread is unlikely to be uniform. It is concentrated in a few big cities and states and its spread is less in rural areas and smaller towns and cities.

The lockdown of India for more than two months helped in limiting the health crisis, but – as in other countries – the complete shutdown of all economic activities except essential services has created an economic crisis and misery for the poor, with massive job losses and rising food insecurity.

The economic shock has been much more severe for India, for two reasons. First, pre-COVID-19, the economy was already slowing down, compounding existing problems of unemployment, low incomes, rural distress, malnutrition, and widespread inequality. Second, India's large informal sector is particularly vulnerable. Out of the national total of 465 million workers, around 91% (422 million) were informal workers in 2017-2018. Lacking regular salaries or incomes, these agricultural, migrant, and other informal workers would be hardest hit during the lockdown period. Here, I focus on the likely impacts on agriculture, supply chains, food and nutrition security, and livelihoods.

Agriculture and supply chains

COVID-19 is disrupting some activities in agriculture and supply chains. Preliminary reports show that the lack of available migrant labor is interrupting some harvesting activities, particularly in northwest India where wheat and pulses were harvested. There are disruptions in supply chains because of transportation problems and other issues. Prices have declined for wheat, vegetables, and other crops, yet consumers are often paying more. Media reports show that the closure of hotels, restaurants, sweet shops, and tea shops during the lockdown is already depressing milk sales. Meanwhile, poultry farmers have been badly hit due to misinformation, particularly on social media, that chickens are carriers of COVID-19.

Some measures required to keep the agricultural sector and supply chains working smoothly are listed here:

1. The government has correctly issued lockdown guidelines that exempt farm operations and supply chains. But implementation problems leading to labor shortages and falling prices should be rectified.

2. Keeping supply chains functioning well is crucial to food security. It should be noted that 2 million to 3 million deaths in the Bengal famine of 1943 were due to food supply disruptions – not a lack of food availability.

3. Farm populations must be protected from the coronavirus to the extent possible by testing for the virus and practicing social distancing.

4. Farmers must have continued access to markets. This can be a mix of private markets and government procurement.

5. Small poultry and dairy farmers need more targeted help, as their pandemic-related input supply and market-access problems are urgent.

6. Farmers and agricultural workers should be included in the government's assistance package and any social protection programs addressing the crisis.

7. As lockdown measures have increased, demand has risen for home delivery of groceries and e-commerce. This trend should be encouraged and promoted.

8. The government should promote trade by avoiding export bans and import restrictions.

Using social safety nets as a bridge between health shock and economic shock

The lockdown has choked off almost all economic activity. In urban areas, this has led to the widespread loss of jobs and incomes for informal workers and the poor. Estimates by the Centre for Monitoring Indian Economy show that unemployment shot up from 8.4% in mid-March to 27% throughout April. In urban areas, unemployment was around 30% in April. There was a loss of 122 million jobs in April compared to the employment level in 2019. Only in the first week of June, the unemployment rate declined to 23%. The shutdown has caused untold misery for informal workers and the poor, who lead precarious lives and face hunger and malnutrition.

The best way to address this urgent need is to use social safety nets extensively to stabilize their lives with food and cash.

The Indian government has responded quickly to the crisis and announced a $22 billion relief package, which includes food and cash transfers. Several state governments have announced their own support packages.

The central government's relief package, called Pradhan Mantri Garib Kalyan Yojana (Prime Minister's plan for well-being of the poor), is aimed at providing safety nets for those hit hardest by the COVID-19 lockdown. However, it is inadequate in the face of the enormous scale of the problem. Nobel Prize economists Esther Duflo and Abhijit Banerji say that the government should have been much bolder with the package's social transfer schemes. The $22 billion in spending is only 0.85% of India's GDP.

In the middle of May, the central government announced a Rs. 20.9 trillion ($279 billion) package – 10% of GDP – covering, among others, agriculture, informal workers, and medium, small, and micro enterprises (MSME). One of the criticisms of the package is that many of its measures are related to credit availability and long-term reforms. These measures could be useful in the medium and long term. The real fiscal stimulus is only 1% of GDP. The poor and vulnerable need immediate help.

Below are some further measures needed in addition to the government package:

- **Food and nutrition security.** Government warehouses are overflowing with 71 million tons of rice and wheat. In order to avoid exclusion errors, it is better to offer universal coverage of distribution in the next few months. Nutrition programs like Integrated Child Development Services (ICDS), mid-day meals, and Anganwadis (rural childcare centers) should continue to work as essential services and provide rations and meals to recipients at home. Eggs can be added to improve nutrition for children and women. Several state governments have started innovative programs to help informal workers and the poor. For example, the Kerala government is providing meals with diversified diets at the doorsteps of households.

- **Cash transfers.** Unemployed informal workers need cash income support. The government has provided Rs.500 ($6.60) per month to the bank accounts of 200 million women via the Jan Dhan financial inclusion program. But this too is insufficient. A minimum of Rs.3000 ($40) per month in cash transfers is needed for the next three months.

- **Migrant workers.** There are about 40 to 50 million seasonal migrant workers in India. In recent days, global media have broadcast images of hundreds of thousands of these migrant workers from several states trudging long distances on highways; some walked more than 1,000 kilometers to return to their home villages. These workers should be given both cash transfers and nutritious food.

COVID-19 is an unprecedented challenge for India; its large population and the economy's dependence on informal labor make lockdowns and other social distancing measures hugely disruptive. The central and state governments have recognized the challenge and responded aggressively – but this response should be just the beginning. India must be prepared to scale it up as events unfold, easing the economic impacts through even greater public program support and policies that keep markets functioning.

Originally published April 8, 2020, and updated June 15, 2020.

DIETS AND NUTRITION

8. The COVID-19 nutrition crisis: What to expect and how to protect

Derek Headey and Marie Ruel

The COVID-19 pandemic has all the makings of a perfect storm for global malnutrition. The crisis will damage the nutritional status of vulnerable groups through multiple mechanisms. We can expect a dangerous decline in dietary quality in low- and middle-income countries (LMICs) stemming from the income losses related to government-mandated shutdowns and de-globalization, as well as from the freezing of food transfer schemes such as school feeding programs and the breakdown of food markets due to both demand shocks and supply constraints. But malnutrition will also increase due to healthcare failures, as already strained healthcare systems are forced to divert resources from a range of nutritionally important functions – including antenatal care, immunization, micronutrient supplementation, and prevention and treatment of childhood diarrhea, infections, and acute malnutrition – toward combating COVID-19.

Based on evidence from previous crises and some limited evidence from the current pandemic, we outline here what to expect and how to protect the most vulnerable, especially women and children, from the effects of this nutritional crisis. We also emphasize the critical need for high-frequency surveillance of vulnerable populations (for example, through phone surveys) and close coordination across sectors, including health, agriculture, education, water and sanitation, social protection, and commerce and trade. Just as in normal times, malnutrition remains a multidimensional problem during times of crisis and therefore requires multisectoral solutions.

COVID-19 may lead to drastic reductions in dietary quality

The COVID-19 economic crisis will affect diets primarily through declining demand for vegetables, fruits, and animal-sourced foods, which are the main sources of essential micronutrients in diets. But these demand shocks will also break down the value chains that supply such highly perishable foods, further exacerbating the shift to monotonous, nutrient-poor diets.

Income effects are likely to be dramatic for poor households in LMICs because of widespread unemployment resulting from COVID-19 mitigation measures. The poor will respond by purchasing the cheapest calories they can find to feed their families. We know from previous IFPRI research that in poor countries calories from nutrient-rich, nonstaple foods like eggs, fruits, and vegetables are often as much as 10 times more expensive than calories from rice, maize, wheat, or cassava. In the face of drastic declines in income, vulnerable households will quickly give up nutrient-rich foods in order to preserve their caloric intake.

This happened during the 1998 Indonesian financial crisis, when real wages fell by 33% between August 1997 and August 1998 due to rising unemployment and a food price crisis. Strikingly, even as rice prices skyrocketed by almost 200%, rice consumption continued to rise during this period. A nutritional surveillance study in rural Java that collected 14 rounds of data during 1995–1997 also found dramatic declines in egg, meat, and vegetable consumption. Not surprisingly, child anemia, sometimes caused by iron and other micronutrient deficiencies, rose sharply from a baseline of 52% to 68% at the peak of the crisis, and children's mean weight-for-height declined by over one-third of a standard deviation.

The COVID-19 economic crisis could also affect nutrition through disruptions to supply chains for nutrient-rich foods. Most nutrient-rich foods are highly perishable, resulting in fragile supply chains. A breakdown in any part of the supply chain – from farms to traders, transporters, and processors to retailers – can break the whole chain. IFPRI has already collected evidence of significant disruptions to livestock production in China and vegetables and dairy value chain disruptions in Ethiopia, while Indian media are reporting a full-blown crisis in dairy marketing, which is particularly troubling in a country where milk and milk products are key sources of essential nutrients for young children. Public food distribution programs are gearing up in several countries to mitigate these types of problems, but such programs deliver nonperishable staples, oils, and pulses, potentially compounding the tendency toward poor-quality diets.

In addition to the severe deterioration in household diets, there are major additional nutritional risks for mothers and young children. Agencies like UNICEF are worried about disruptions to imports of crucial nutritional products, including micronutrient supplements and micronutrient-fortified products used to prevent and treat micronutrient deficiencies or severe acute malnutrition. Anecdotal evidence from Asia also suggests that women are concerned about passing the coronavirus to their infants through breast milk, which could result in switching to breast-milk substitutes that increase risks of infection and malnutrition in areas with poor water quality. And globally, breastfeeding promotion and nutrition counseling usually provided by the health sector will be severely limited due to restrictions on mobility, social distancing requirements, and overburdened healthcare systems.

COVID-19 threatens maternal and child health, directly and indirectly

Pregnant women and mothers with young children are obviously vulnerable to COVID-19, and especially so if they have other underlying health conditions. But the indirect effects on healthcare systems will likely have far greater consequences for maternal and child health. In principle, lockdown protocols in most countries do not prohibit health-related travel, but healthcare providers and their clients will be less willing to travel for non-emergency check-ups or preventive care (such as immunization, nutrition counseling, and micronutrient supplement distribution), and many maternal and child healthcare providers have already been reassigned to COVID-19 responsibilities.

The erosion of resources for maternal and child healthcare could easily manifest in invisible tragedy, characterized by drastic declines in antenatal, neonatal, and essential maternal, infant, and child healthcare services. Some of the most direct risks include breakdowns in supply and/or distribution of antenatal iron folic acid or multiple micronutrients; child vitamin A supplementation; distribution of oral rehydration salts and zinc for diarrhea and therapeutic food for home treatment of acute malnutrition; and safe in-person treatment consultations.

More indirect, yet equally severe, is the risk that diversion of healthcare resources toward combating COVID-19 will jeopardize regular but life-saving efforts to prevent and treat malaria, diarrhea, and other infectious and tropical diseases. In much of Asia and Africa, the inevitable scaling back of these efforts will coincide with the monsoon rains – a time when, even in normal years, the incidence of tropical infectious diseases and of acute malnutrition rises steeply. In 2020, the perfect storm of eroded basic healthcare, chronic lack of access to safe water, sanitation, and hygiene (WASH), declining dietary quality, and heightened seasonal risk of infectious disease means that many millions of children in Asia and Africa will be in danger of severe, life-threatening disease and malnutrition. Poor nutrition, in turn, weakens the immune system and can jeopardize the body's ability to fight a COVID-19 infection.

How can we protect vulnerable groups?

Policymakers and researchers alike are operating in a unique state of uncertainty, and those working on nutrition may struggle to get their voices heard in the fog of this war against COVID-19. It is therefore critical to strengthen and broaden multisectoral nutrition coalitions to ensure that actors in different sectors work as effectively as possible to prevent a full-blown nutritional crisis. All over the world, COVID-19 response committees have been formed to address the crisis, but in many instances, we fear, those championing food and nutrition security are being sidelined. Yet it is now more critical than ever that multisectoral nutrition groups advocate and support key actions to protect nutritionally vulnerable groups (with many of these actions also contributing to poverty reduction). These include:

1. **Keep agrifood systems functioning.** Let farmers farm, traders trade, input dealers deal, and sellers sell, even if they are informal. Implement social distancing and improve hygiene measures along the value chain, but keep domestic and international food markets working.

2. **Facilitate food system innovations.** Given that social distancing and mobility restrictions may be in place for many months, governments, development partners, and microfinance institutions should search for ways to stimulate innovative and safe food delivery systems, for example, especially those that create jobs.

3. **Support local (or homestead) food production to increase access to nutrient-rich vegetables, fruits, and eggs and improve diet quality.** These programs are consistent with social distancing, can use surplus household labor, including women, and will increase consumption of nutrient-rich foods.

4. **Find innovative ways to stimulate demand for nutrient-rich foods.** National leaders and national media must urge their populations to keep consuming healthy diets. Mobile phone messaging could be used to stimulate demand for protective nutrient-rich foods and to encourage appropriate infant and young child feeding practices, including optimal breastfeeding and diet diversity practices.

5. **Use social safety net programs to improve dietary quality, not just quantity.** Food transfers are often focused on staples, and where available, should consider biofortified (micronutrient-rich) crops. Cash transfers or vouchers schemes linked to innovative food delivery systems

should also be considered to keep the economy going and stimulate demand for fruits and vegetables, dairy, and other nutrient-rich foods. School feeding programs should also adopt new modalities to safely distribute food during school closures.

6. **Prevent the collapse of basic maternal and child health services.** The medical profession is facing the most challenging crisis of the past century. Still it must find ways to maintain basic preventive and curative healthcare services, especially for mothers and young children, whether through remote consultations and nutrition counseling (using mobile messaging or radio) or COVID-19–safe home visits and delivery of essential drugs and supplements. Key to meeting this challenge is increasing access to protective equipment for all healthcare workers.

7. **Invest in WASH, urgently.** WASH programs are a win-win for preventing contagion of COVID-19 and other infectious diseases that affect maternal and child health and nutrition. Many LMICs have increased access to hand-washing stations in communal places. Public announcements and mobile messaging can raise awareness and nudge individuals into more hygienic routines.

8. **Ramp up support to community-based management of acute malnutrition.** With the expected rises in acute malnutrition, it will be important to boost (safe) community-level screening and referral of children with acute malnutrition, maintain appropriate stocks of life-saving supplements, and ensure appropriate staffing and availability of protective equipment. Regular monitoring and surveillance will also be needed to assess the emergence of acute malnutrition among newly vulnerable populations.

9. **Protect women and children.** Economic stress and social distancing will increase the risk of domestic violence and psychosocial stress. Social protection and other relief programs need to prioritize women and children and explore novel ways to support individuals and communities in the context of prolonged social distancing.

10. **Set up or scale up food and nutrition surveillance systems.** These systems help in identifying the scope and scale of nutritional crises, especially fast-moving crises. Innovations in phone and web-based surveillance systems offer new tools for timely monitoring of vulnerable populations to improve targeting and program design in a time of unparalleled uncertainty.

The CGIAR Research Program on Agriculture for Nutrition and Health (A4NH), led by IFPRI, provided support for the production of this blog.

To read more about this topic, see: D. Headey, R. A. Heidkamp, S. Osendarp, M. T. Ruel, N. Scott, R. Black, H. Bouis, et al. "Impacts of COVID-19 on Childhood Malnutrition and Nutrition-Related Mortality." Lancet. Article in press. First published online on July 27, 2020.

Originally published April 23, 2020.

9. COVID-19 is shifting consumption and disrupting dairy value chains in Ethiopia

Agajie Tesfaye, Yetimwork Habte, and Bart Minten

The COVID-19 crisis is having a range of impacts on food consumption and value chains everywhere – containment measures, lost incomes, and perceptions of disease risk are altering food availability and consumer preferences. To understand the effects of the COVID-19 crisis on Ethiopia's important dairy sector, we conducted a qualitative appraisal of the dairy value chain supplying Addis Ababa. Between April 15 and May 10, we interviewed nearly 100 commercial and small dairy farmers in urban and rural areas, dairy processors, traders, development agents, urban retailers, and consumers.

Overall, the survey indicates that the Ethiopian dairy sector has experienced only moderate impacts – especially compared to the livestock sectors in China and other countries. However, certain segments of the industry – particularly raw milk vendors and small dairy shops – have been hit hard.

Downstream: Urban retail and consumption

Data from two recent large-scale household surveys indicate that the consumption of dairy products in Addis Ababa has decreased since the start of the COVID-19 crisis. In January/February 2020, 56% of residents questioned said that they consumed dairy products in the previous seven days. In May, this number declined to 45% of interviewed households (Figure 1). All income groups decreased their consumption, except for the richest quintile, where the share of consuming households changed little.

An important reason for falling consumption is the fear of disease risk. More than half of respondents in the household survey said they were avoiding the consumption of animal-sourced foods (meat, milk, yogurt, cheese) due to the perceived COVID-19 risk. This widespread perception is apparently linked to Ethiopian media reports at the beginning of the outbreak suggesting that consumption of fish and animal-sourced products was associated with greater chances of infection. However, COVID-19 transmission via food is not considered a significant problem, and the risk from dairy products in particular is minimal; the virus typically spreads through close person-to-person contact via airborne respiratory droplets.

While media outlets later corrected these warnings regarding dairy products, the Ministry of Health has warned the public to avoid consumption of raw foods because of the potential risk of contamination through droplets coming from food handlers. Thus the perception of risk from dairy products remains – particularly from raw milk. Our interviews indicate a significant drop in the demand for raw milk; a steady, or even higher, consumption of pasteurized milk; and an increase in purchases of powdered milk, as the latter two are considered safer by consumers.

FIGURE 1 Dairy consumption in Addis Ababa, Jan./Feb. and May 2020

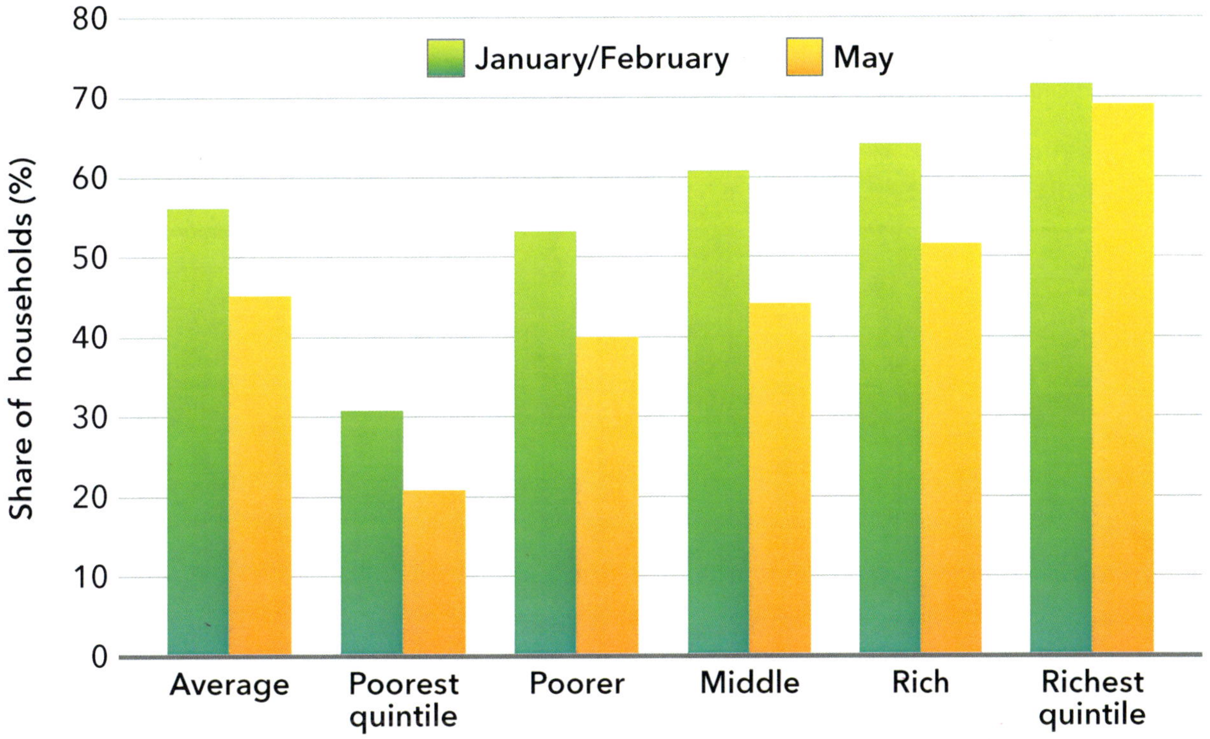

Source: Wolle et al. (2020); Hirvonen et al. (2020).

Addis Ababa's dairy shops – where people can buy milk and yogurt, and there is also often space to sit and eat – are estimated to distribute 11% of the city's liquid milk, much of it in raw form, and more than 25% of its yogurt. Dairy shop owners or managers we interviewed all complained of a sharp reduction in customers, largely because consumers link raw milk, or yogurt made from raw milk, with an increased risk of contracting the virus.

Social distancing measures are also impacting business. Dairy shops are often located in areas that are now much less busy – such as near universities, whose students have gone home. Most are also small and cannot easily accommodate orders to keep customers widely separated. Such distancing measures are also leading to a drop in demand for milk products by coffeehouses and pastry shops, we find, another important outlet for dairy.

We also see a decrease in activities by small informal distributors of raw milk. Supplied by urban dairy farms and also in part by small-scale milk collectors, they sell their product to urban residents in plastic jerrycans of 10 to 20 liters. They often also use public transport. The consumer-clients whom we talked to indicated that they were scared of buying from such vendors due to perceived COVID-19 risks. The reasons mentioned included the high number of visits such traders make to different houses, their lack of health precautions, and the fear of contamination of utensils used by collectors, milkmen, or vendors across this raw milk marketing chain.

Interviews with owners of regular grocery shops – which are very important for the distribution of dairy products in Addis – and of supermarkets indicate that demand was down, as usual, during the fasting period in March and April, and that it has since returned to normal levels, or that it was even up compared to the same period in other years. Some indicated that they were running out of supplies. But dairy demand has shifted: They all indicated that the demand for powdered milk, which normally makes up almost 10% of the dairy expenditures of urban households in Addis, significantly increased. They suggested that consumers believe that the processed product is less risky than raw milk and is also appealing because it can be stored indefinitely – important given the uncertainty surrounding stay-at-home measures, the risk associated with going out, and the fear that food supply chains might break down.

Midstream and upstream: Dairy processing companies, collectors, and farmers

These changes in consumer preferences pass down through the dairy value chain – in particular, distributors, collectors, and rural farmers involved in the raw milk value chain have been severely affected by the COVID-19 crisis. Farmers are less able to sell their milk. As a result, more milk was processed and the butter supply rose, and butter prices fell sharply in rural areas. Some dairy households reported incidences of wasted milk as they could not find buyers, and some indicated they were increasing their own dairy consumption.

Meanwhile, prices of liquid milk remained stable in urban retail markets and prices for producers supplying dairy processing plants did not change. Marketing margins did not change very much either. On the cost side, feed prices increased 30% to 40% over several months, before dropping back to normal levels. Wheat bran and oil-cake were the most affected. The spikes may have been due to wheat factories receiving less supply from rural areas and/or reducing operations by placing some workers on leave in response to COVID-19. Transportation problems may also have contributed. But overall, all farmers interviewed said production has not fallen since the start of the COVID-19 crisis.

We also assessed access to veterinary medicines. These seem to be less available in public pharmacies but can still be found in private ones. However, stakeholders in rural production areas indicated that prices had gone up by 15% to 20% since the start of the COVID-19 crisis. Retail prices often increased because prices from distributors had increased, as their international supply channels might have been disrupted.

Conclusion

Overall, despite the exceptions, the Ethiopian dairy sector has proven surprisingly resilient in the face of the pandemic, control measures, and consumer worries about food contamination. This might be explained by the fact that the Ethiopian production system is much less dependent on marketed inputs than that of some other countries. Nevertheless, our survey indicates a number of problems remain, including the challenge of accurately communicating the COVID-19 disease risks associated with food to consumers. It is also important to further monitor these developments. A new household survey that will be fielded in the coming weeks in Addis Ababa might provide new needed insights.

We thank Anne Bossuyt (IFPRI), Fantu Bachewe (IFPRI), Kaleab Baye (Addis Ababa University), and Rinus van Klinken (SNV) for comments on an earlier version of this post.

This work was funded in whole or part by the United States Agency for International Development (USAID) Bureau for Food Security under Agreement # AID-OAA-L-15-00003 as part of the Feed the Future Innovation Lab for Livestock Systems, implemented by the Institute of Food and Agricultural Sciences of the University of Florida in partnership with the International Livestock Research Institute (ILRI).

Originally published June 1, 2020.

10. Survey suggests rising risk of food and nutrition insecurity in Addis Ababa, Ethiopia, as COVID-19 restrictions continue

Kalle Hirvonen, Gashaw Tadesse Abate, and Alan de Brauw

As the COVID-19 pandemic has spread to virtually every corner of the world, lockdowns, supply disruptions, and economic pain have followed in its wake, raising alarm about food and nutrition security among policymakers, the development community, and other observers. Representative survey data on households' immediate and longer-term responses to the pandemic are necessary to understand these impacts and their implications, and to plan and target appropriate responses. The need is especially urgent for urban areas, where residents may face greater public health risks and tighter restrictions.

To cast light on how households in Addis Ababa, Ethiopia, are reacting to the crisis, IFPRI's Ethiopia Strategy Support Program (ESSP), with the support of the CGIAR Research Program on Agriculture for Nutrition and Health (A4NH), has begun a series of monthly phone surveys. Initial data demonstrate that poorer households are taking a greater economic hit than those with higher incomes, and that dietary diversity has declined. The results suggest the food security situation in Addis Ababa could sharply deteriorate in the coming weeks if disease transmission and social distancing measures continue.

The phone surveys build on data collected from a representative sample of households that participated in a randomized controlled trial in 2019, with the endline taking place in early 2020. The first phone survey was conducted in early May and covered 600 households, as will subsequent rounds, with an emphasis on ensuring household respondents are spread across the Addis Ababa income distribution.

Ethiopia confirmed its first COVID-19 case on March 13. The Ministry of Health immediately began contact tracing and isolating those who tested positive for the virus. Three days later, the government closed schools, banned all public gatherings and sporting activities, and recommended social distancing. Other measures to prevent the spread of the virus soon followed. Travelers from abroad were put into a 14-day mandatory quarantine, bars were closed until further notice, and travel across land borders was prohibited. Several regional governments imposed restrictions on public transportation and other vehicle movement between cities and rural areas.

While the policies were clear, as in many other countries there was confusion, mixed implementation, and a range of economic fallout. Our initial phone survey, conducted between May 1 and May 6, 2020, aimed to capture the immediate effects on people and their knowledge of the disease. It shows

FIGURE 1 Change in income levels in April 2020 compared to usual incomes, by household wealth quintile, Addis Ababa

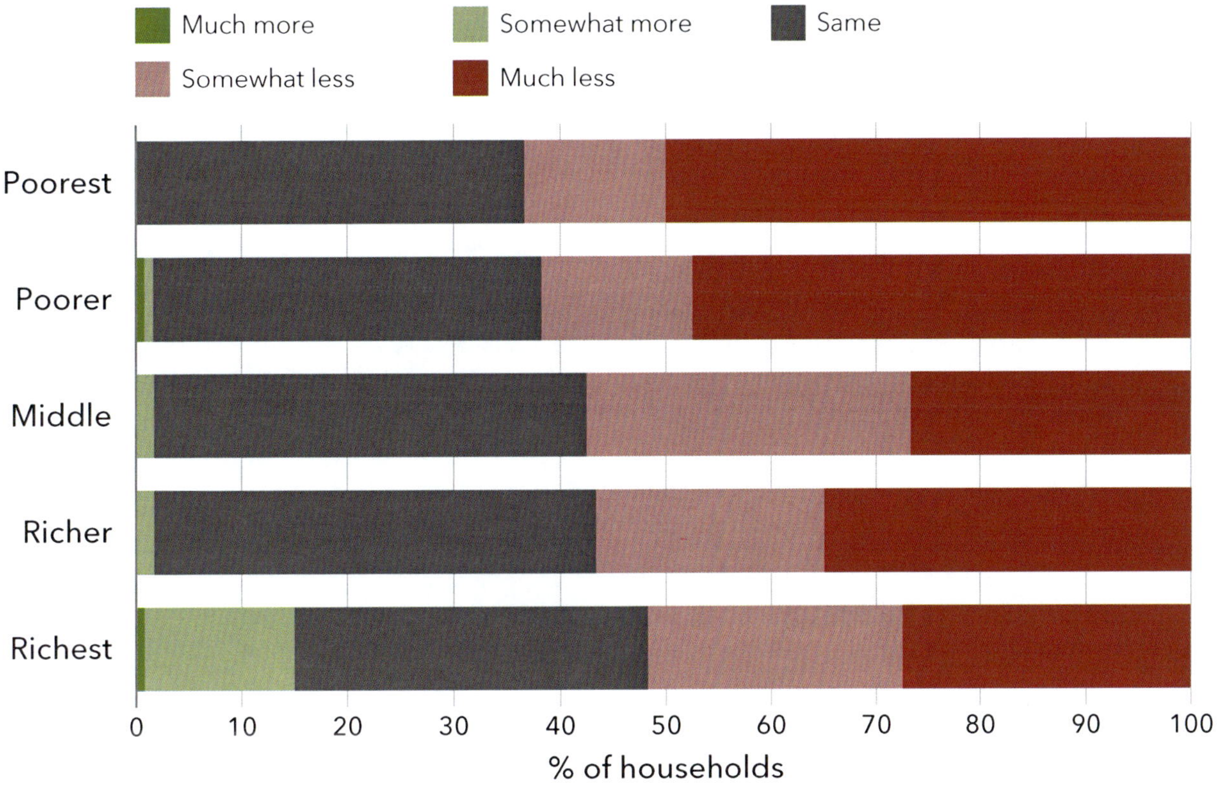

N = 600 households

Source: Authors' calculations from Addis Ababa COVID-19 phone surveys.

that almost all households are aware of COVID-19, and most are aware of basic preventative measures such as hand-washing and social distancing. Like many people in developed countries, a relatively large share of the sample – 35% – stated that they are "extremely stressed."

We asked respondents to compare their household income in April 2020 to their usual household income at this time of year (Figure 1). About 37% of respondents stated their households had "much less income" and another 21% stated their households had "somewhat less income." When we examine reported losses by wealth quintile (defined using the survey conducted in January and February), we find poorer households are much more likely to report much less income during April than richer households.

How are households dealing with reduced incomes? We asked those who reported income losses to share their primary strategy, and we find that about two-fifths report using savings, and another one-fourth report reducing their household food expenditures. Households do not report having much savings on average; about two-fifths have only enough savings for up to 14 days of food expenditures, and only 10% of all households report having enough savings for more than a month of food expenditures.

FIGURE 2 Changes in household reports of consumption of specific foods between February and April 2020, Addis Ababa

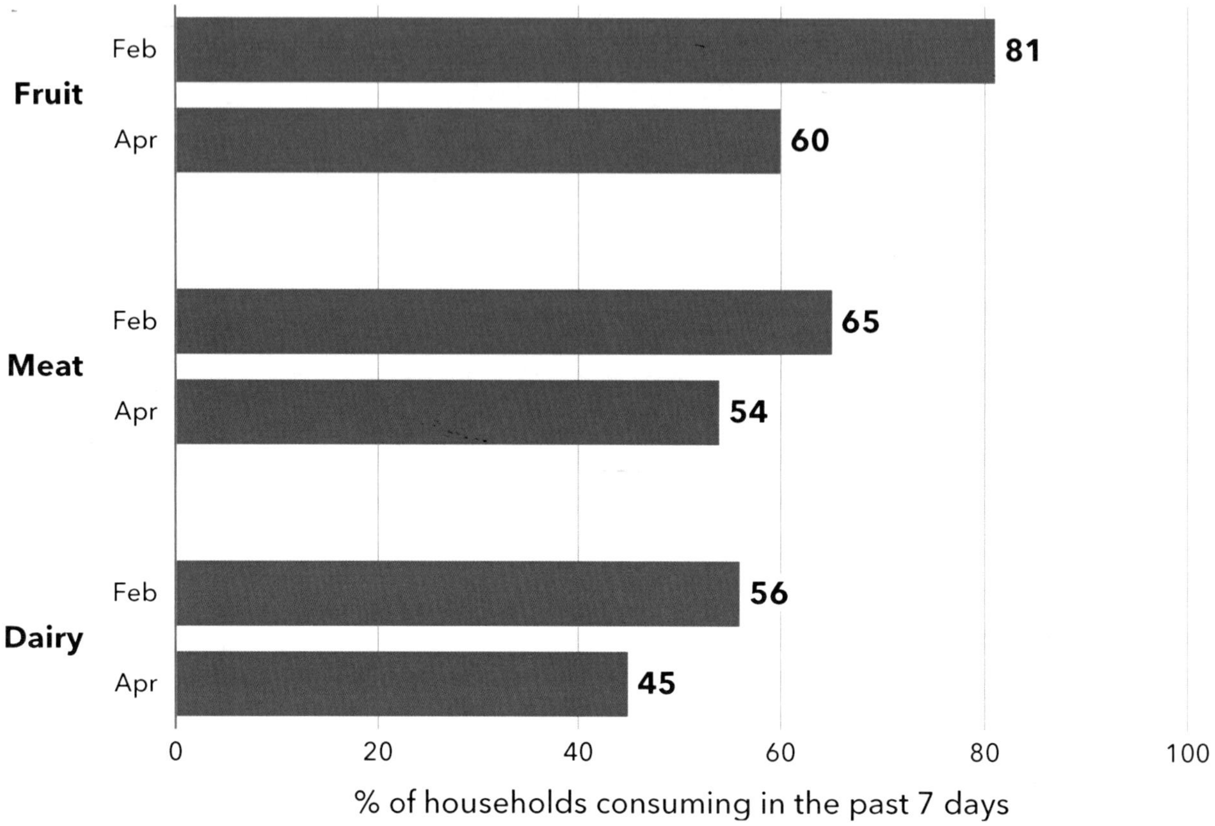

N = 600 households

Source: Authors' calculations from Addis Ababa COVID-19 phone surveys.

With declining food expenditures, we observe diet changes already beginning to occur. We administered a standard household dietary diversity score module, repeated from the (in-person) survey in early 2020. We find households are less likely to report consuming fruits (declining from 81% to 60% of households), meat (65% to 54%), and dairy (56% to 45%) (Figure 2). These results suggest concerns about the declining nutrient density of diets are real: households are reducing food expenditures by substituting away from more nutrient-dense foods to cheaper, less-nutrient-dense foods. These results are worrisome, as there is substantial uncertainty about how long the pandemic and its economic impacts will persist, and in the long term such dietary changes could both increase malnutrition and be detrimental to the food system as a whole.

A major disruption like a pandemic has serious implications for the evolution of food systems. When negative shocks occur, food purchase decisions can shift from a focus on dietary diversity toward ensuring there is enough to eat. Our data show that many households in Addis Ababa have done just that, though their overall food security status is not yet alarming. However, as personal savings dwindle, the likelihood that we will observe a rapid increase in food insecurity in the near future is quite high if COVID-19 restrictions continue.

While these measures limit the spread of COVID-19, they come at a high cost, particularly to the poorest households. One policy response would be to rapidly scale up existing support programs before the food insecurity and hunger situation reaches alarming levels. In urban Ethiopia, the Urban Productive Safety Net already provides an established framework for identifying the poorest and most affected households. Another important response is to support the 1,000 food banks that have been established in Addis Ababa to curb the likely deterioration in food security. Only concerted assistance can help to sustain a population cut off from its income and facing fewer, and worse, food choices as a result. In further survey rounds, we plan to continue tracking both how policies reach sample households and how food and nutrition security change among these households.

The CGIAR Research Program on Agriculture for Nutrition and Health (A4NH), led by IFPRI, provided support for the production of this blog.

Originally published May 21, 2020.

LABOR RESTRICTIONS AND REMITTANCES

11. Lockdowns are protecting China's rural families from COVID-19, but the economic burden is heavy

Scott Rozelle, Heather Rahimi, Huan Wang, and Eve Dill

In response to the COVID-19 outbreak in December 2019, China implemented a nationwide travel blockade and quarantine policy that required all public spaces, businesses, and schools to shut their doors until further notice and placed restrictions on individuals leaving their homes or traveling.

The lockdown was also implemented across China's vast rural areas, home to more than 700 million people. These quarantine measures started during the annual Spring Festival in mid-January, when most rural residents had returned to their family homes to celebrate the Lunar New Year together. Many were migrant workers who had expected to return to China's urban and industrial centers to continue working in factories, construction sites, and service sectors.

In China's urban hubs, local governments, school systems, and businesses made efforts to offset the consequences of these policies: many firms worked with employees remotely through online platforms, and urban schools moved to online learning activities. These efforts helped reduce the fears and economic repercussions for those who were able to work from home and had access to high-speed internet. But what about the rest of China – the "Other China"?

Little is known about what actions were taken in rural areas as part of the nationwide quarantine, and even less is known about the effects of COVID-19 in China's rural villages outside the COVID-19 epicenter during and after the quarantine. To date, no study has empirically examined the economic and social impacts of COVID-19 or its countermeasures in a rural context. China's rural residents are a relatively poor subpopulation, with a meager social safety net at best. Understanding the economic and social effects of COVID-19 on China's vulnerable rural population can offer urgently needed lessons as the outbreak spreads to other middle- and low-income countries and regions around the world.

To assess the effects of COVID-19 control measures on the health and economy of China's rural population, a team of researchers, led by the Rural Education Action Program (REAP) at Stanford University, conducted phone surveys with 726 randomly selected village informants across seven rural Chinese provinces outside of Hubei, the epicenter of the virus.[1]

1 The response rate was almost 100%, since the respondents were either the mother or the father of a rural student who had been part of a study that REAP had conducted over the past year or two. After answering the phone, our survey enumerators identified themselves as a team member who had been a part of their child's school activities in the recent past. Parents almost summarily agreed to talk to us. The schools in the original studies were randomly chosen and the one parent of the child from each school was randomly chosen, meaning we have a sample that is fairly representative of rural areas outside of the epicenter.

Our village-level survey examined three overarching questions:

1. What disease control measures were in place?

2. How many COVID-19 infections and fatalities were there in each village? (That is, were the dramatic quarantine measures working to stop the virus from spreading?)

3. What were the indirect impacts of these disease control measures on employment, health (beyond COVID-19 issues), and schooling?

First and foremost, our survey in mid-February found that all villages had universally implemented extremely strict quarantine measures to stop the spread of the virus:

- 100% of all villages had erected strict and high barriers to quarantine their villages off from the rest of China.

- 98% reported that all group gatherings (including weddings and funerals) had been temporarily banned.

- 97% reported that villagers could not visit the homes of friends or family within the village.

- 86% reported that even their close family or friends living outside the village were not permitted to enter.

- 96% of informants reported that villagers were required to wear masks to go outside (although only 16% reported that masks were available for purchase).

- 95% reported they could leave the village to seek healthcare.

So, what was the direct impact on the spread of COVID-19?

This is the good news: The survey was clear that the draconian quarantine measures successfully contained the spread of COVID-19 in rural villages. Only 4 village informants out of 726 reported COVID-19 infections in their villages, and of the nearly 700,000 residents in these villages, only about 10 had contracted the virus. No one in any surveyed village reported deaths from the virus. This suggests that lockdown measures can effectively minimize the spread of the virus.

However, the question remains: what is the cost in terms of the lives and livelihoods of rural villagers?

Most notably, we found virtually no one was working in the off-farm sector – either in a city as a migrant worker or in the local township/county seat as an off-farm laborer. Three-quarters of informants reported that villagers had stopped working because their workplaces were closed. An even greater share could not work due to restrictions on transportation or difficulty in finding housing in the places they typically worked. This means that the employment of rural workers was essentially zero for a full month after the start of the quarantine. Not surprisingly, 92% of village informants reported that disease control measures had reduced their income levels.

Our research also found a number of other impacts from the lockdowns on education, nutrition, and access to healthcare:

- 79% reported a negative impact on local children's education.
- 63% reported that the prices of foodstuffs were higher than in 2019. Although the majority said fruits, vegetables, and grains were all available, nearly half said the quality of their diets fell – raising questions about the impact on nutrition.
- 62% believed it had become more difficult to seek non-COVID-19 healthcare.

While everyone in China was feeling the effect of the COVID-19 outbreak after one month of restrictions, it is almost certain that rural residents took the brunt of the economic and social impacts. Our analysis suggests a radical decline in employment in China's rural areas, due at least in part to restrictions preventing migrant workers from returning to work. Whereas urban, salary-earning workers were getting paid during the quarantine as required by edicts issued by China's central government, rural workers are almost never on salary – if they don't work, they don't get paid. If we conservatively assume 75% of rural migrants were confined to their villages in February, nearly 200 million rural individuals, who make an average of $500 every month, were not working. On a larger scale, this means that after one month of COVID-19 restrictions, China's economy lost around $100 billion in rural migrant worker wages alone. If we then add the lost wages of the large rural workforce that live and work near their home villages, the total economic loss is significantly higher than $100 billion and exceeds the highest estimate of the global economic impact of SARS – and it still does not account for all the other losses to the economy.

At the same time, however, there have been some positive developments. As in urban areas, rural governments have taken measures to reduce the negative effects of COVID-19 by encouraging online schooling: 71% of village informants reported that students were attending classes online. However, we have yet to determine the quality and rate of their learning in online classes. Our follow-up study will tell us more.

Now we're left asking, what happens after all control measures are lifted and rural residents are forced to try to provide for their families with a significant loss in income?

Our team conducted a follow-up phone call survey in mid-March, which looks predominantly at how people reacted to their economic losses when the quarantine policies were ending. Even with the lifting of the restrictions on movement in March, at least half – and potentially up to 60% or 70% – of the rural workers, who had been working in the previous year, were still not working. The radical decline in employment during and after the quarantine clearly was already impacting the livelihoods of rural communities. Over half (53%) of the villages surveyed reported their local workers had lost approximately two months' worth of income. This represents about 17% of their annual income. As a result, families have been forced to decide what essential commodities to cut down on so as to survive on their now-limited funds. Villagers in the survey villages reduced their spending on food (55%), education (10%), and (non-COVID-19) healthcare (9%). The prices of common goods in 2020 also were reported to be higher than the previous year (2019) in both February (63%) and March (66%) surveys. In practice, this means that people are buying more grains and staples in bulk at low cost in lieu of more expensive goods like meat and produce. It also indicates that nutrition has declined – at least among a share of rural families. This is particularly concerning for families with young children, as REAP's past research shows that nutritional deficiencies in early childhood can significantly inhibit cognitive development, which is linked to adverse outcomes in later life.

As COVID-19 continues to spread across the globe, our findings have strong implications for other countries that have adopted similar lockdown policies. Workers around the world are facing potentially huge losses of income in the coming weeks and months. As governments implement COVID-19 control measures, they must also consider the needs of economically vulnerable communities, or face dramatic increases in economic hardship and poverty among the hardest hit.

Originally published March 30, 2020, and updated June 15, 2020.

12. Economic impact of COVID-19 on tourism and remittances: Insights from Egypt

Clemens Breisinger, Abla Abdel Latif, Mariam Raouf, and Manfred Wiebelt

The economic impacts of the COVID-19 crisis are increasingly hitting low- and middle-income countries and the poor. International travel restrictions and the full or partial closure of businesses and industries in Asia, Europe, and North America have led to a collapse in global travel and are expected to reduce the flows of remittances. Tourism and remittances are important sources of employment and incomes for the poor. This post assesses the potential impacts of the expected reductions in these income flows by using Egypt as a case study.

The pandemic is likely to have a significant economic toll. For each month that the COVID19 crisis persists, our simulations using IFPRI's social accounting matrix (SAM) multiplier model for Egypt suggest national GDP could fall by between 0.7% and 0.8% (EGP 36-41 billion or US$2.3-$2.6 billion). Household incomes are likely to fall, particularly among the poor.

Egypt is a rising star among emerging economies. Even though several reforms remain to be completed, the reform program launched in 2016 has started to bear fruit: Egypt has achieved economic growth of over 5% in the last two years. The tourism sector recorded its highest revenues in 2018-19, another sign of increased stability. Continued efforts aimed at improving Egypt's business climate were expected to lead to even stronger private sector growth and economic diversification in 2020 and beyond.

This progress will almost certainly be interrupted by the COVID19 pandemic. While the government is taking actions to contain the spread of the virus – including the suspension of commercial international passenger flights, school and sports clubs closures, and a nationwide nighttime curfew – and the number of reported infections in Egypt is currently low compared to that of many other countries, the global economic slowdown is expected to have major knock-on effects for Egypt. International travel restrictions are already curtailing tourism to the country. The global slowdown is likely reducing payments received from the Suez Canal and remittances from Egyptians working abroad. These three sources together account for 14.5% of Egypt's GDP. Thus, any disruptions to these foreign income sources will have far-reaching implications for Egypt's economy and population.

Using the SAM multiplier model for Egypt, we simulate the individual and combined effects of a collapse in the tourism sector and reductions in Suez Canal revenues and in foreign remittances under more and less pessimistic scenarios. SAM multiplier models are well-suited to measuring short-term direct and indirect impacts of unanticipated, rapid-onset demand- or supply-side economic shocks such as those caused by the COVID-19 pandemic. We model the demand shocks as the anticipated reductions in tourism, Suez Canal, and remittances revenues.

FIGURE 1 Estimated GDP loss per month, less pessimistic and more pessimistic scenarios, as percentage of average 2019 monthly GDP

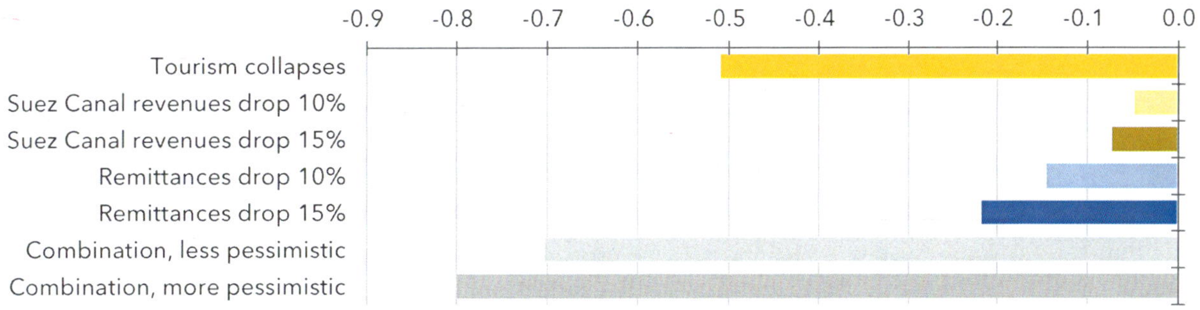

Source: Authors' calculations.

Note: The less pessimistic combination scenario assumes a 10% reduction in Suez Canal revenues and in remittances. The more pessimistic scenario assumes a 15% reduction in these payments. Both combination scenarios assume a complete absence of international tourists. GDP = gross somestic product.

Our results show the potential significant impact on the economy and people for each month that the COVID-19 crisis persists. If the dynamic effects of the COVID-19 shock on the Egyptian economy are different than those simulated, our results could be either under- or over-estimations of the aggregate economic impact of the crisis. Also, effects from other channels may reinforce the effects of the pandemic.

These expectations also assume that there is no change in the current government policies in place to combat the crisis. This is important to note, as the government is taking aggressive action to contain the disease and support the economy and people. As such, the model scenarios do not consider the impacts of specific government economic policies, but are intended to support government decision-makers in determining the scale of their support to the economy and to Egyptian households.

Figure 1 breaks down estimated losses in GDP, which may hit 0.8% per month in the more pessimistic scenario. Lower tourist spending will affect not only hotels, restaurants, taxi enterprises, and tourist guides, but also food processing and agriculture. Lower public revenues from Suez Canal fees are likely to affect the government budget. Lower remittances income will likely reduce household consumption of consumer goods and hit sectors producing intermediate goods. We estimate that the absence of tourists alone may cause monthly losses of EGP 26.3 billion, or $1.5 billion. Thus, the loss in tourism revenues accounts for about two-thirds of the total estimated impact.

Household incomes are estimated to decline by between EGP 153 or $9.70 (less pessimistic scenario) and EGP 180 or $11.40 (more pessimistic scenario), per person per month for each month that the crisis continues (between 9.0% and 10.6% of household income). The expected reduction in tourism has the strongest effect on all households, making up more than half the economic impact for all household types in the model (Figure 2). Households are also affected directly and indirectly by lower remittances from abroad.

FIGURE 2 Estimated household consumption loss per month under the less pessimistic scenario, disaggregated by source of loss

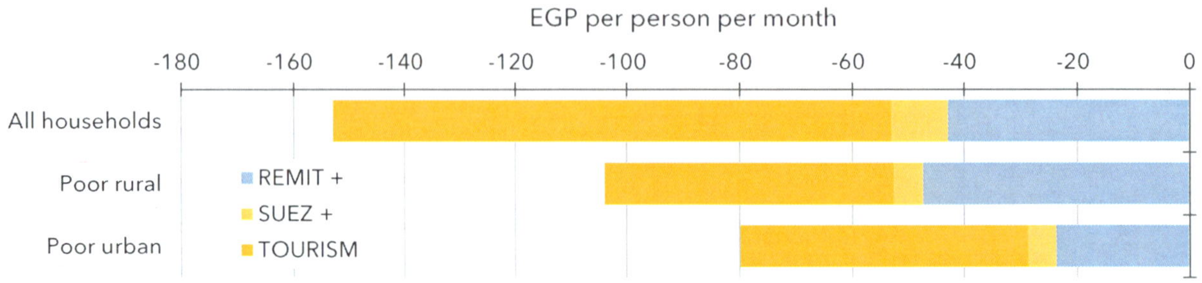

Source: Authors' calculations.

Note: The less pessimistic scenario assumes a 10% reduction in Suez Canal revenues and in remittances. EGP = Egyptian pounds.

While all households are hurt by lower tourist expenditures, it is poor households – and especially those in rural areas – that suffer the most from lower remittances. Due principally to the relatively greater decline in remittances that they experience, rural poor households are estimated to lose in total between EGP 104 and 130 ($6.60-$8.20) per person per month, or between 11.5% and 14.4% of their average income, while urban poor households will see their incomes decline somewhat less, between EGP 80 and 94 ($5-$6) per person a month, or between 9.7% and 11.5% of their average income.

Policy considerations

If the crisis persists for at least three to six months, as many now believe likely, the cumulative loss from these three external shocks alone could amount to between 2.1% and 4.8% of GDP in 2020. Importantly, our simulations measure only the effects that might result from specific impact channels, namely, foreign sources of remittances and revenues. Domestically, restrictions on movement of people and goods within the country and on certain productive activities may also have adverse economic impacts. On the other hand, some sectors may benefit, such as information and communications technologies, food delivery, or the health-related goods and services sectors.

The authorities have begun a course of decisive action to curb the virus outbreak by allocating EGP 100 billion ($6.3 billion) and have enacted tax breaks for industrial and tourism businesses, reducing the cost of electricity and natural gas to industries, and cutting interest rates. They are also considering providing grants to seasonal workers. Additional measures may also be in the works, such as increasing cash transfer payments to poor households, increasing unemployment benefits, and providing targeted support to specific sectors.

While the country's focus currently is rightly on fighting the health crisis and mitigating its immediate impacts, planning on how to re-open the economy should start now. To emerge stronger after the COVID-19 crisis, both the public and private sectors should continue to strengthen their collaboration. The government should work to further improve the business climate for the private sector and continue undertaking serious reforms to overcome institutional weaknesses. The crisis may also provide an opportunity to strengthen analytical capacity in Egypt to provide policymakers with research-based solutions for safeguarding Egypt's economy during future pandemics and other crises.

Unless governments around the world take decisive action, the case of Egypt suggests that poverty is likely to increase in countries where tourism and remittances play a large role. It is also a strong reminder of the interconnectedness of the world and the importance of global cooperation to end this crisis and to be better prepared for the future.

We thank Dr. Diaa Noureldin, Senior Advisor; Dr. Sahar Aboud, Principal Economist; Racha Seif Eldin, Senior Economist; and Mohamed Hosny, Economist, all at the Egyptian Center for Economic Studies (ECES), for their inputs to this study through their excellent work on ECES' Views on News - Views on the Crisis series. We also are grateful to Xinshen Diao, James Thurlow, and Karl Pauw, all of IFPRI, for their technical review and comments.

A more detailed description of the SAM multiplier model and the underlying assumptions are published as an IFPRI Policy Note.

Originally published April 1, 2020.

13. Significant economic impacts due to COVID-19 and falling remittances in Myanmar

Xinshen Diao and Michael Wang

The COVID-19 pandemic and government lockdown in Myanmar have led to falling exports and lost revenue from tourism and international remittances, hitting the economy hard. In a new series of policy notes, we examine the economic impacts of the pandemic and restrictive measures to mitigate the health crisis, and offer policy recommendations to address declining incomes and other impacts.

Our analysis shows a major short-term economic contraction as a result of the two-week lockdown in April – a 41% decline in GDP along with similar declines in most nonagricultural sectors in comparison to the same period without a pandemic. This is not surprising, as Myanmar's economy is deeply integrated into a complex supply network both domestically and internationally, and policies affecting certain industries have ripple effects on other sectors through supply and demand linkages. In addition, approximately 4 million Myanmar migrants work internationally, and their lost income due to lockdowns in neighboring countries is expected to impose ongoing significant burdens on low-income households that receive remittances.

In our analysis, we applied a social accounting matrix (SAM) multiplier model to evaluate COVID-19's direct and indirect effects on Myanmar's economy. The SAM multiplier model is a simulation tool that describes the economic connections between national economic actors and provides a highly disaggregated picture of the economy, which is suitable for measuring the impacts of short-term shocks.

The lockdown and subsequent restrictive measures have had direct and indirect negative impacts on the flow of goods and services, resulting in a decline of 41% in national GDP during the lockdown (Figure 1). The figure breaks down the decline further to show the different impacts of COVID-19 restrictions on Myanmar's various economic sectors.

We estimate that agricultural GDP fell 14% during the two-week lockdown. While agricultural activities were mostly exempt from restrictions, the linkages with sectors in the rest of the economy led to significant indirect impacts, including reductions in demand from non-agricultural sectors, falling exports, lower consumer demand from falling remittance income, and difficulties in operating agribusinesses.

On their own, these negative short-term economic shocks are sufficiently large to temporarily push many Myanmar households into poverty and food insecurity. Moreover, our analysis shows that the sharp decline in remittance income is likely to continue for at least a year if not longer. As a result, many low-income households that sit just above the poverty line are expected to fall into poverty, and the extremity of poor households will increase (Figure 2).

FIGURE 1 Estimated percentage change in Myanmar's GDP during the April 2020 two-week lockdown period by sector, compared with a normal situation without COVID-19 in the same period

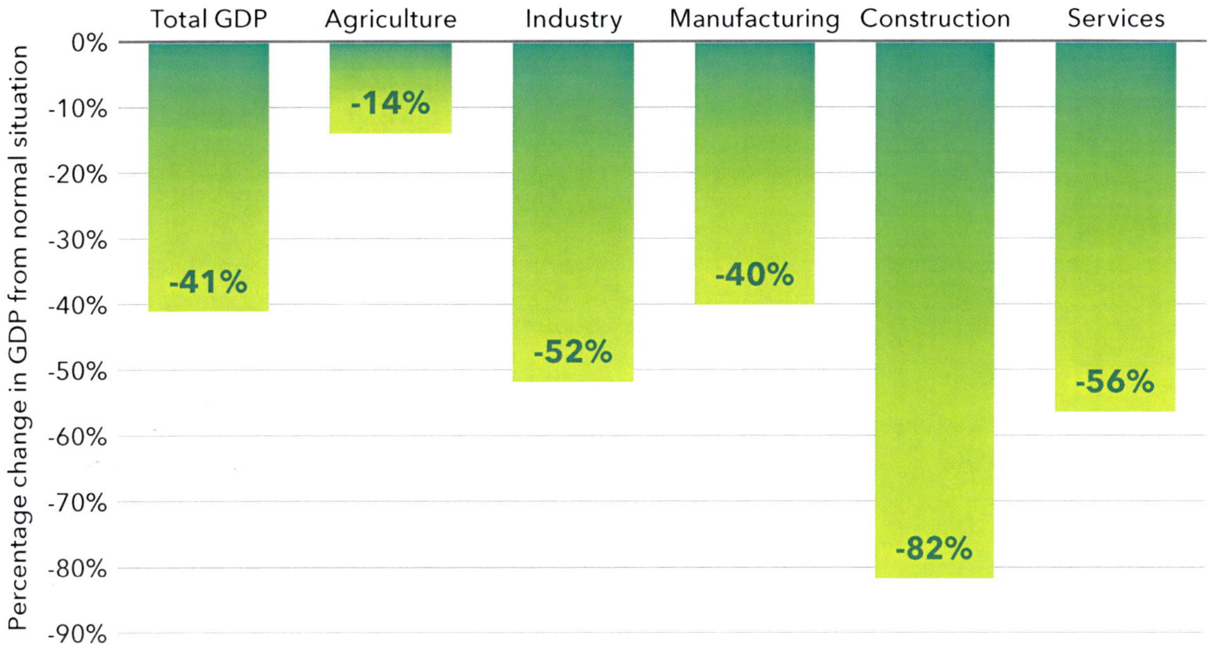

Source: IFPRI social accounting matrix (SAM) model.

FIGURE 2 Percent declines in total income due to a 50% international remittance shock and a 30% domestic remittance shock among poor and low-income remittance-receiving households

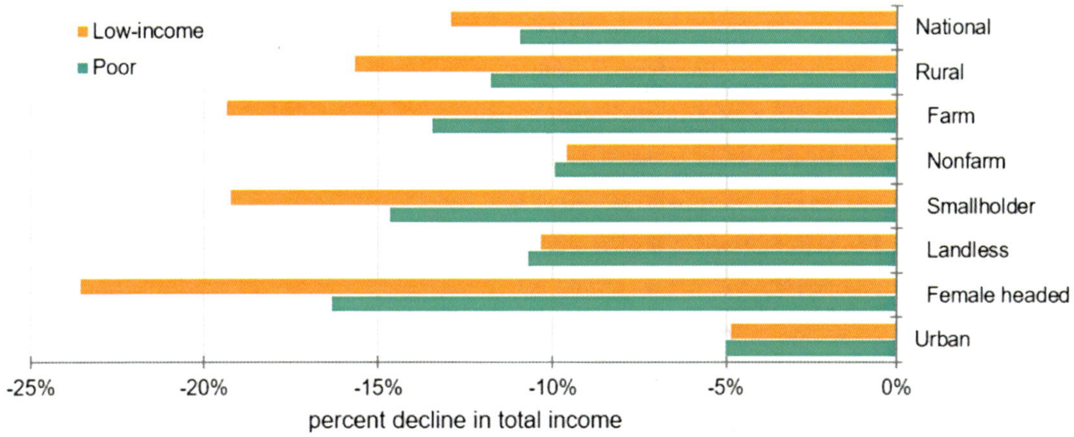

Source: SAM model.

Note: Farm, nonfarm, smallholder, landless, and female-headed households are part of rural households. Includes remittance-receiving households only.

Declines in total income from remittance shocks are consistently higher among low-income rural households than poor rural households. However, though the shocks might result in smaller relative income losses for poor rural households, these households will see significant impacts because their income levels were initially much lower. Note that negative impacts on total income are largest for female-headed rural households.

The government recently released a comprehensive and sensible economic response package, the Myanmar COVID-19 Economic Relief Plan (CERP), which includes unconditional cash and in-kind transfers to the most vulnerable and affected households. The current analysis could be helpful in identifying potential recipients among remittance-receiving households and in determining the amount of financial support they need. The anticipated total spending under CERP will be around 2.8 trillion kyat (about US$2 billion). Considering that the loss in national GDP estimated in our analysis is between 6.4 trillion and 9.0 trillion kyat by the end of FY 2020, the size of this economic stimulus package might be too modest to enable all firms, households, and the whole economy to return to their pre-COVID-19 growth trajectories in 2021.

This chapter is based on a policy note prepared at the request of the Myanmar Agricultural Policy Support Activity (MAPSA). Funding for the study was provided by the CGIAR Research Program on Policies, Institutions, and Markets (PIM), led by IFPRI, and United States Agency for International Development (USAID).

Originally published June 25, 2020.

FOOD TRADE

14. COVID-19: Trade restrictions are worst possible response to safeguard food security

Joseph Glauber, David Laborde, Will Martin, and Rob Vos

As COVID-19 spreads around the globe, fears of a deep global recession are mounting. Some also fear that food supplies may start running short, especially if supply chains are disrupted. Others fear that agricultural production may be disrupted by containment measures that restrict workers from harvesting and handling crops.

While we should take these concerns seriously – especially for fruits and vegetables, which have complex supply chains, or foods sold primarily through restaurants – they should not be overstated either, especially not for basic staples such as rice, wheat, and maize. Global markets are well supplied, stocks are healthy, production of key staples is unlikely to be disrupted, and prices have remained relatively stable. Trade is allowing production to move from areas of surplus to areas of shortage, avoiding the drastic shortages and food insecurity associated with reliance only on local production.

But there will be serious threats to poor people's access to food as a consequence of lost income from lockdowns and other restrictions. These should be addressed through measures that help maintain access to food, rather than through policies like export bans that may further threaten that access.

The food price crisis of 2007-2008 shows, however, that policy concerns about food availability can easily turn into a serious price crisis. At the time, some grain-exporting countries responded by imposing export restrictions, which pushed up world market prices of staples, leading other grain producers to also limit exports in efforts to insulate their consumers from the initial food price rises. Food-importing countries, worried about the higher cost of food, in turn lowered import tariffs on food, supporting demand but keeping upward pressure on world prices. As a result, instead of containing price increases, these policy responses only drove world market prices higher. In the case of rice, these policy responses contributed almost half of the world price surge in 2007-2008.

Unfortunately, once again several countries responded by implementing export restrictions, though fortunately many of these were temporary. See our online tracker. Kazakhstan, for instance, suspended exports of several cereal products, as well as oilseeds and vegetables, until June 30. Viet Nam halted granting rice export certificates through the end of March, but has since begun granting them again. These restrictions, even if temporary, seem entirely unnecessary. Both countries produce far more than they consume and have ample stocks. An export ban by two key exporters would limit global supply and will certainly push up world prices of staple foods if others follow suit.

How does the present situation compare with 2007-2008? Will we see a repeat of the same policy mistakes?

FIGURE 1 Staple crop stock-to-use ratios, 2008 and 2019

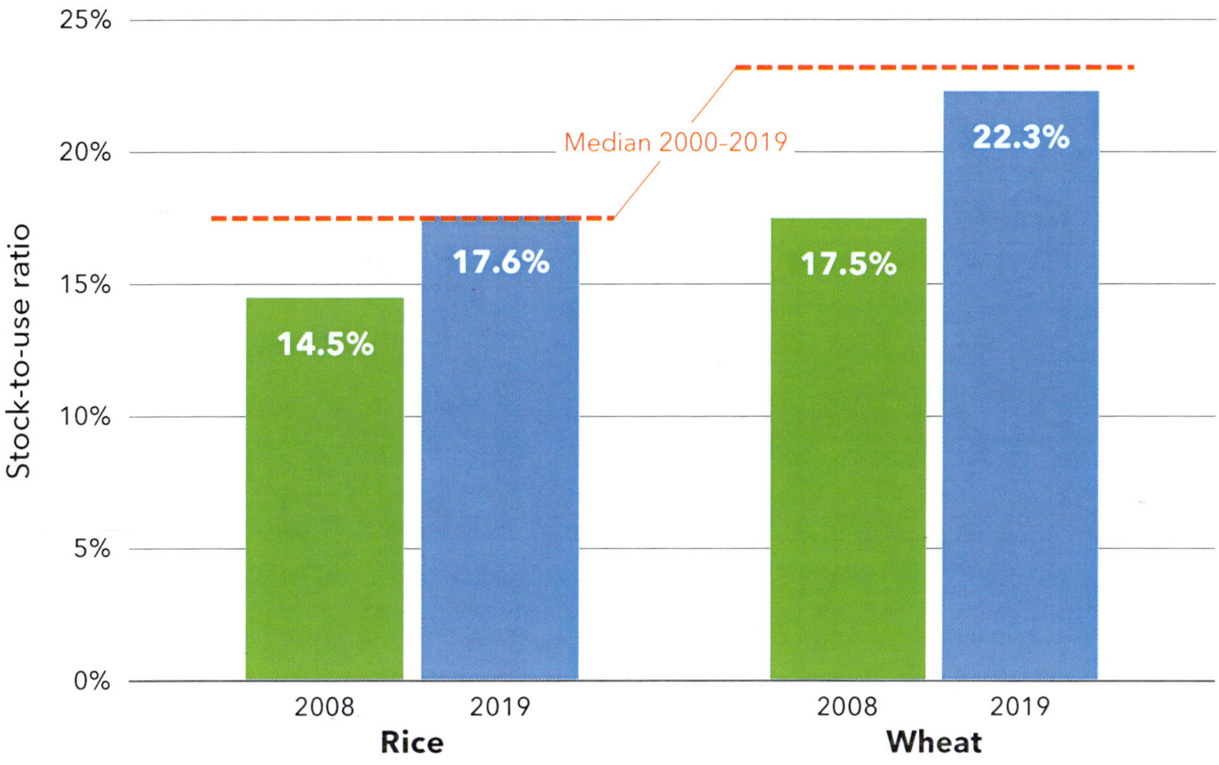

Source: Authors' computation using USDA-PSD data.

Some key facts

There is no shortage of staple food inventories. The stock-to-use ratio is a critical indicator of the vulnerability of world food markets to shocks. Excluding China, the current global stock-to-use ratios are close to their "normal value" (the median level of the last two decades), and substantially higher than in 2008 (Figure 1), when markets were tight. The sufficiency of inventories explains in good part the relative price stability in the markets for staples. The underlying situation is better than suggested by these statistics when also considering China's inventories of rice and wheat, which are sufficient for 10 to 13 months of domestic consumption.

Harvests are expected to be good. The US Department of Agriculture projects an increase in global wheat production of 5%, while rice production is projected to remain about the same as in 2019. Production of these key staples is unlikely to suffer disruptions from the COVID-19 crisis – at least in major producing countries – since much of it is mechanized, requiring relatively little labor input, and takes place in areas with dispersed, already socially distanced, rural populations. Similarly, there is low probability of disruptions to international transport and distribution of these key staples which, being dry bulk commodities, can be loaded, shipped, and discharged with minimum human-to-human interaction.

World exports are heavily concentrated. Russia, the European Union, the United States, Canada, and Ukraine together are likely to account for 75% of all wheat exports in 2019-2020. It therefore matters a great deal what governments of these countries do. So far, only Kazakhstan, which has a 3% share in global wheat exports, has announced export restrictions. However, Russia is now also reportedly considering a ban on wheat exports. The rice market is equally concentrated, with 75% of exports coming from the largest five exporters, and nearly a quarter from India alone. Viet Nam's world market share is 16%, and as noted above it has suspended new export licenses. India's stock-to-use ratio for rice, however, stands at an historic high of 34% and prospects for the 2020 harvest are good, such that it should have no reason to consider export restrictions, although some concerns have been expressed about difficulties moving products domestically.

What should be done?

The present outlook for staple food markets is much brighter than it was during the 2007-2008 price spike. Hence, imposing trade restrictions now would be even more misguided than it was in 2008. Rather, such policies could become the problem if Viet Nam and Kazakhstan maintain barriers and other countries follow in their footsteps. If they do, it could trigger food price spikes and speculative behavior in agricultural commodity markets. The world's poor would be the ones bearing the brunt.

Instead, trade channels should be kept open so that international markets can play an instrumental role in in avoiding food shortages and mitigating the inevitable global economic downturn.

Major exporters and importers of staple foods should agree to desist from imposing trade barriers in response to the COVID-19 pandemic. Rather, as we wrote in a previous blog post, the focus should be on measures that will help stave off a global recession and minimize a further rise in food insecurity that way. For this, governments will need to provide fiscal stimulus, including resources to contain the spread of the disease and ensure adequate healthcare is available, as well as additional social protection to compensate workers and families affected by the virus and by containment measures.

This work was supported by the CGIAR Research Program on Policies, Institutions, and Markets (PIM), led by IFPRI, and United States Agency for International Development (USAID).

Originally published March 27, 2020.

15. COVID-19 border policies create problems for African trade and economic pain for communities

Antoine Bouët and David Laborde

The COVID-19 pandemic has triggered a range of border controls in countries around the world to curb the spread of the disease. In Africa, these moves have interrupted progress toward economic integration. The African Continental Free Trade Area (AfCFTA), for example, was supposed to establish continentwide free movement of goods starting on July 1. Now, the African Union Commission has proposed postponing the launch until January 1, 2021. In addition, trade restrictions implemented in Africa and elsewhere in response to the pandemic are fueling fears of a new food crisis on the continent (see IFPRI's tracking of export restrictions).

Across Africa, pandemic-related border controls are having many economic impacts, large and small. Here, we examine these impacts and suggest ways to soften the blow to affected people and communities.

Most African countries have closed land borders to travelers, while still allowing freight to pass under tighter controls, which sometimes allows the movement of only agricultural and food products. Over one 11-day period in March, 24 African countries imposed such measures on land borders (Figure 1). Almost all these countries have also suspended the arrival of international flights, at least from countries particularly affected by the virus. Many governments have also imposed curfews.

The Democratic Republic of the Congo, Kenya, Liberia, and Namibia chose a different path: the entry of people at border posts is subject to temperature control and testing, followed by hospitalization and/or quarantine if necessary.

These measures have been adopted to protect public health, but their economic consequences could be significant. Stricter sanitary border controls on the transport of products are likely to slow intra-African trade. In addition, prohibiting people from crossing borders stops one means of informal trade, widely practiced in Africa and often the main source of income for a family. Informal trade accounts for a significant share of recorded trade: for example, between 15% and 30% of official exports in Uganda.

The consequences of these measures for intracontinental trade are still unclear due to a lack of recent data. Thus far, statistics compiled by the Food Security and Nutrition Working Group (weekly data collected at border posts in East Africa – the only data available through the end of March) do not indicate a decrease in cross-border agricultural trade. Most of the border closures took place in the second half of March, so it is too early to tell if data capture any effects. And only five countries (Djibouti, Ethiopia, Rwanda, Sudan, Uganda) implemented border closures in East Africa during this period.

FIGURE 1 Closure of land borders in Africa, March 2020

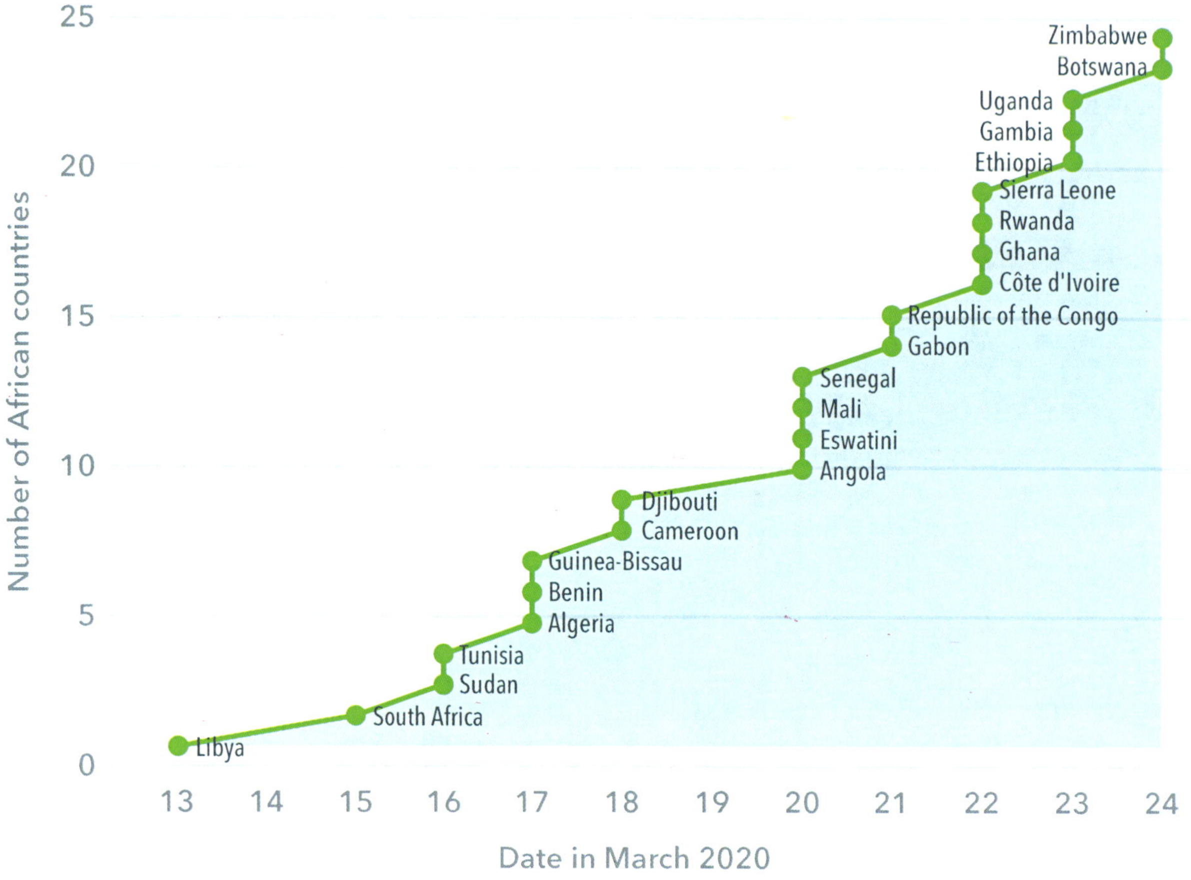

Source: Authors' elaboration from websites of US embassies in Africa and from al Jazeera.

Problems with border policies

Most border closures have been imposed with little clear knowledge of what is happening on the ground. For example, in West Africa, because of daytime heat, fresh produce, meat, and other perishable products are usually transported at night. Yet curfews make this practice impossible. Mandating more thorough health checks without adding necessary personnel also increases transport times. Health check delays and curfews are likely to cause significant waste and loss of products in West Africa, according to an interview with Brahima Cissé, a trade analyst with the Permanent Interstate Committee for Drought Control in the Sahel (CILSS).

Border restrictions on travel can be particularly costly for livestock producers practicing transhumance – seasonally moving livestock between grazing grounds. This occurs between Sahelian countries including Burkina Faso, Mali, and others to coastal countries such as Benin and Côte d'Ivoire; and between Kenya and Uganda. Beyond their immediate economic costs, these measures threaten the basic mode of operation of pastoral agriculture.

Travel restrictions can also make access to inputs such as fertilizers or pesticides more difficult.

The introduction of exceptional measures provides a breeding ground for the abuse of power. In many parts of Africa, it is common practice for law enforcement officials to set up checkpoints along trade corridors to collect bribes. As recent measures have slowed road transport in West Africa, this predatory behavior has increased in intensity: according to Cissé, bribe collection has increased by 30% per truck along these corridors since March.

Most of these measures were imposed with little warning, taking local populations by surprise and leaving them to contend with the fallout. With informal trade interrupted, many people have had little opportunity to find alternative livelihoods. For many families, the absence of income for even several consecutive days can have devastating effects on poverty and food security.

There has also been little international or regional coordination of these border-related decisions. For example, curfew times often vary between neighboring countries, compounding their economic impacts.

Finally, such measures may interrupt international technical assistance (health and/or food aid), imposing significant economic, public health, and other costs.

Potential solutions

To address these hardships, governments should provide ample safety nets to those affected, for example, informal traders making their living from cross-border trade. But safety nets are costly and difficult to design. How to set up these transfers in a period of confinement (especially in the absence of possible digitalization of payments in some countries)? How can measures be put in place that take into account the specific vulnerability and role of women?

The World Health Organization (WHO) has often expressed reservations about the border crossing bans and their role in protecting public health. They increase the likelihood that people will cross borders through places not covered by customs authorities and evade health checks. The WHO is also concerned that governments might avoid publicly acknowledging an outbreak in order to avoid having their citizens targeted by other countries' trade and travel restrictions.

Border checkpoints should be set up to provide health checks and screening, possibly followed by quarantine and/or hospitalization for the infected. Such a system can provide important health information to the population and improve the distribution of protective equipment, soap and disinfection equipment, and access to water. In the East African Community (EAC: Burundi, Kenya, Rwanda, South Sudan, Tanzania, Uganda), nine mobile laboratories have recently been deployed to provide systematic testing, particularly along the northern border between Uganda and Kenya.

Physical distancing requirements at border crossings may also reduce the spread of the virus. But this of course requires supplementing the teams of customs officers working at border stations, so as not to slow down cross-border trade too much.

To reduce the costs for farmers and transporters of agricultural and food products, governments should reconsider curfews, which hurt the transport of perishable products. In terms of intra-African trade policy, import taxes on agricultural and food products should be reduced to compensate for higher transport costs. A suspension of export bans on these same products should also be considered.

New border restriction measures should be announced in advance in order to allow people to adapt as best as possible. Countries should also coordinate their policies to allow for exchanges of information on the spread of the virus and responses. The WHO Africa Regional Office and the Inter-African Bureau for Animal Resources can help in this regard. Regional Economic Communities can also play an important role. In addition to the EAC, the Economic Community of West African States (ECOWAS) is studying a plan of action including the lifting of all land border and port restrictions on the free movement of agricultural inputs, including fertilizers and pesticides, and the promotion of social safety net projects for food and nutrition.

Finally, countries should not let the pandemic stop progress toward economic integration. The need for the AfCFTA has been reaffirmed by influential figures such as Presidents Paul Kagame of Rwanda and Cyril Ramaphosa of South Africa, as it can provide not only a solid basis for long-term economic development, but also a means of effectively fighting future pandemics by facilitating the cross-border trade of food and medical goods. Virtual negotiations could begin in the coming days to set a new start date, possibly before January 1.

Defining coherent policies in health and economic terms in the face of a pandemic such as COVID-19 is a particularly complicated exercise. It already seems to be very difficult in rich countries with significant financial resources and strong institutions. It is obviously even more difficult in poor countries, where financial resources are very limited and institutions are sometimes weak. Policies adapted for countries with strong institutions can be inappropriate, or even harmful, in countries with weaker ones. For example, as we have seen, imposing stricter health controls along trade corridors can increase the predatory behavior of local control authorities and make the situation worse. The international community must therefore help these countries to take into account the institutional environment when implementing these policies.

Originally published May 12, 2020.

16. COVID-19 lockdowns threaten Africa's vital informal urban food trade

Danielle Resnick

As COVID-19 begins its spread across Africa, concerns are growing about how the pandemic will affect the region's already fragile food systems, especially in densely packed cities.

Much of the region's urban population works in the informal sector – many in wet markets and as street vendors – and depends on it for food, so lockdowns and other social distancing measures could pose major problems both for consumers and workers.

Traders often migrate daily to city centers on minibuses and via other forms of public transportation, work in very close proximity to each other, and do not have the ability to take off work if they feel sick. National lockdowns, such as those in Rwanda, South Africa, and Zimbabwe, and city lockdowns in Benin, Côte d'Ivoire, Democratic Republic of the Congo, Ghana, Nigeria, and Uganda, could prove disastrous, since such traders provide the majority of food to Africa's urban poor. How food traders are managed could have substantial ripple effects on the nutrition and income prospects of many across Africa.

How will governments respond as the pandemic continues to spread? The record is worrying.

First, African governments have a history of cracking down on informal traders, especially during public health crises. When the Zambian government used the military to close down markets during Lusaka's 2018 cholera outbreak, farmers who sold their fresh produce to informal traders lost a significant amount of income.

Second, as the trend of government decentralization has widened in Africa, many of these markets and street vending activities have fallen under the mandates of local governments. Thus they now generate significant tax revenue – not only for critical public services to combat COVID-19, such as water, health clinics, and waste collection, but also to pay the salaries of local government bureaucrats. Shutting them down would therefore have negative effects on the broader urban political economy.

Third, during the food price crisis of 2007–2008, African cities were major sites of unrest, and this could reoccur if prolonged market lockdowns cause a large spike in food prices. There are already some early signs of price increases for food staples in Rwanda and in Kinshasa. Spiking food prices can spark protests, which would present yet another public health problem: in a region with relatively low levels of state capacity, it could be difficult for governments to peacefully discourage group protests to avoid the spread of infection.

To manage this potential looming crisis in urban centers, local political leaders should be communicating now with market leaders about how to best handle a possible shutdown. Despite outward appearances of disorderliness, many markets are actually well-governed by cooperatives or associations organized along product lines. Their leaders could identify what supplies traders need to stay safe, where to set up hand-washing stations, and ways to reduce density by alternating the days traders come.

To both discourage travel to markets and still provide traders with some income, city governments could also consider temporarily relaxing bylaws that prevent citizens from selling outside their homes. Based on India's experience with a nationwide lockdown thus far, some other feasible options include opening markets every other day and sanitizing on the off days, and allowing for trading around the clock to reduce consumer congestion.

In the longer term, crises represent critical junctures for essential reforms and innovations. Every year, thousands die due to cholera outbreaks in African cities because of unsafe water and sanitation, and markets can be major sites of infection. At the extreme, Zimbabwe's capital, Harare, shut down its main waterworks in late 2019 due to a lack of foreign exchange to import treatment chemicals, deepening an ongoing national water crisis. If washing hands with clean water and soap are the main way to prevent the spread of COVID-19, perhaps the pandemic will finally lead to larger investments in this crucial area.

Similarly, like their counterparts elsewhere in the world, informal food traders in Africa have long been excluded from traditional safety nets enjoyed by those in the formal sector, including sick leave and pensions. Due to their diverse backgrounds and volatile incomes, they can also be insufficiently targeted by cash transfer programs that rely on means or proxy-testing. Instead, traders usually rely on rotating credit groups and neighborhood and funeral associations to offer support in the event of idiosyncratic shocks. With a systemic shock like COVID-19, such ad hoc social coping mechanisms are likely to be strained. Thus, it's key to find ways to make social protection systems more inclusive of diverse sets of urban traders. In this regard, South Africa's promise to create a safety net for informal workers in response to this crisis should be closely watched.

Informal food traders in Africa are the heartbeat of their food systems, providing income for farmers, nutrition for poor consumers, and resources for essential urban services. It is imperative that African governments account for them in their COVID-19 responses, and not further relegate them to the shadows of the economy.

Originally published March 31, 2020.

FARAMAREEN
Fly

SUPPLY CHAINS

17. How COVID-19 may disrupt food supply chains in developing countries

Thomas Reardon, Marc F. Bellemare, and David Zilberman

COVID-19 is spreading through the developing world. Many low- and middle-income countries are now reporting growing numbers of cases and imposing rigorous lockdown regulations in response, which impact all aspects of the economy. How will COVID-19 affect food supply chains (FSCs) in developing countries?

The evidence suggests that the impacts will be felt widely, but unevenly. Farm operations may be spared the worst, while small and medium-sized enterprises (SMEs) in urban areas will face significant problems. Governments will have to develop policies to respond to these varied impacts to avoid supply chain disruptions, higher food prices, and severe economic fallout for millions of employees.

For context, here is what the literature tells us about FSCs in developing countries (see also Table 1):

- Most urban and rural consumers now depend on markets, in contrast to 30 to 40 years ago when a large share of rural populations lived "off the grid" in subsistence agriculture. Consumers purchase 80% of all food consumed in Africa and Asia, and thus FSCs provide 80% of all food consumed (Reardon et al. 2019).

- Modern FSCs (dominated by large processing firms and supermarkets, capital-intensive, with relatively low labor-intensity of operations) constitute roughly 30%-50% of the food systems in China, Latin America, and Southeast Asia, and 20% of the food systems in Africa and South Asia.

- Transitional FSCs (stretching from rural to urban areas, fragmented and dominated by thousands of labor-intensive SMEs) dominate food systems, constituting 50%-80% of the food economies of developing Asia and Africa. SMEs in transitional FSCs in developing countries tend to be found in clusters such as dense sets of food processing SMEs, scores of meal vendors at truck stops, and dense masses of wholesalers and retailers in public wholesale markets and wet markets. Each of these clusters could have numerous SMEs. In these venues, large numbers of clients gather in dense crowds.

What might happen to food supply chains

Here are seven hypotheses, based on what we know so far, about the likely effects of COVID-19 on FSCs in developing regions:

1. **Direct impacts will overwhelmingly be felt post-farm.** Namely, in the "midstream" (e.g., wholesale, logistics, and processing) and "downstream," in food-service enterprises.

2. **The impacts are likely to be largest in dense urban and rural peri-urban areas.** Given the properties of the novel coronavirus, which is transmitted most easily via human contact, greater population densities tend to facilitate its spread.

TABLE 1 The three stages of food supply chains and their prevalence in the food economy

	TRADITIONAL FSC	TRANSITIONAL FSC	MODERN FSC
Approximate prevalence in Africa & South Asia as share of food economy	10%	70%	20%
Approximate prevalence in Southeast Asia & Latin America as share of food economy	5%	50%	45%
Main enterprise type	Home microenterprise	SMEs, wet markets	Supermarkets, large processors
Length	Short, local	Long, rural-urban	Long, rural-urban, international
Use of arrangements	No contracts, no standards	No contracts, public standards	Emerging contracts, private standards
Technology	Labor-intensive	Labor-intensive	Capital-intensive

Source: Authors.

Note: FSC = food supply chains. SME = small and medium-sized enterprises.

3. **Effects will be strongest in the downstream segments of retail and food service.** These downstream firms are mostly informal-sector SMEs, and are thus labor-intensive with high densities of workers in small spaces. They have little control over the hygiene practices of their product suppliers or their customers' habits.

4. **Retail and food service firms in modern FSCs face fewer problems.** They are far less vulnerable to mandatory business closures, and also face a lower risk of clients and employees contracting the disease. The least affected are likely to be supermarket chains. Their stores can enforce the flow of entering customers and social distancing measures. Supermarkets and fast-food chains also have more control over the food safety and hygienic practices of their FSCs, as they typically vertically coordinate with contracts and private standards (Swinnen and Maertens 2007).

5. **Direct impacts on farm populations and farm production will be much smaller than on the FSC downstream and midstream.** This is because most small farmers in developing countries rely on family labor. The farm sector, however, will be affected indirectly by COVID-19 through the disruption of input supply chains, and of consumer demand due to lost income and other economic impacts of the pandemic.

6. **COVID-19 is likely to increase food prices, both as a cause and consequence of food shortages.** Restrictions on FSC logistics will increase transaction costs and thus consumer prices. Speculative hoarding may occur and trigger price increases. Higher food prices are, in turn, likely to signal impending shortages. These effects can compound each other in a vicious cycle likely to cause social unrest (Bellemare 2015).

7. **COVID-19 responses will create economic hardship.** Enforcing social distancing and limits on internal and external logistics in FSCs will transform health-risk problems into income and employment risks and political risks.

Implications, strategies, policies

Clearly, the segments of FSCs in the developing world most vulnerable to COVID-19 impacts are the midstream and downstream segments. This will present significant challenges for the people working in them and likely lead to broader economic and operational changes going forward.

As most of these FSCs are in the "transitional" stage, they are composed mostly of informal sector SMEs, with employees lacking formal registration and safety nets such as unemployment insurance.

In the short term, millions of these businesses will face lower foot traffic, lower incomes, and substantial unemployment.

In the medium term, COVID-19 impacts on these segments may be like episodes of avian flu in Southeast Asia in the 2000s, which induced rapid concentration, leading to the rise of large processing firms and supermarkets.

How should governments respond to minimize supply chain disruptions and fallout from lockdowns and other restrictions? The general strategy must be two-pronged: implement robust public health measures to slow the spread of disease; and address food security impacts, particularly the potentially enormous effects on income and employment.

This strategy presents significant challenges for developing countries. Addressing the FSC issues will require three complementary policy paths. In the short run, implement new, broad safety nets for SMEs and workers in the midstream and downstream segments of FSCs; for example, governments could use cash-for-work schemes to employ workers to distribute emergency food rations, upgrade sanitation in wholesale markets and wet markets, and maintain essential operations in their own enterprises so that the latter are there when the crisis passes. In the short and medium term, monitor and regulate wholesale markets, retail wet markets, and processing clusters more strictly, and redesign their sites for improved health practices. Finally, make long-term investments to help SMEs change hygiene practices and improve site design to help them remain competitive.

Thomas Reardon thanks the United States Agency for International Development (USAID) under: (1) the Feed the Future Innovation Lab for Food Security Policy Research, Capacity, and Influence (PRCI), and (2) the Feed the Future Sustainable Intensification Innovation Lab (SIIL). Marc Bellemare thanks the National Institute of Food and Agriculture for grant MIN-14-061. David Zilberman thanks the National Science Foundation, CBET, for grant no. 2030362.

Originally published April 2, 2020.

18. Impacts of the COVID-19 crisis on vegetable value chains in Ethiopia

Seneshaw Tamru, Kalle Hirvonen, and Bart Minten

On March 13, the first COVID-19 case was confirmed in Ethiopia. Three days later, the government closed schools, banned all public gatherings and sporting activities, and recommended social distancing. Other measures to contain the spread of the virus soon followed. Travelers from abroad were put into a 14-day mandatory quarantine, bars were closed until further notice, and travel through land borders was prohibited. Several regional governments banned all public transportation and imposed restrictions on other vehicle movement between cities and rural areas.

While these actions are expected to slow the spread of the disease, they are likely to have substantial effects on food value chains, and thus on the livelihoods of farmers and other workers, and on consumption.

To understand these effects, we conducted a qualitative and rapid appraisal of the vegetable value chain. Building on a large value chain survey that IFPRI undertook in February 2020, we conducted phone interviews (March 23–April 2) with key stakeholders along the vegetable value chain from the main producing areas in the Central Rift Valley to Addis Ababa. Small-scale farmers, large-scale investors, brokers, agro-input dealers, and developmental agents were interviewed. Given that this assessment was based on a limited and nonrepresentative number of interviews, caution is warranted for extrapolation of our observations. They should be seen more as hypotheses of impacts on these value chains. (We intend to substantiate these findings with more representative surveys in the near future.)

Effects downstream and midstream in the vegetable value chain

1. **Vegetable trade and consumption are reduced.** There is less trading activity in Addis's vegetable wholesale market (Atkilt Tera) since the start of the COVID-19 crisis, despite this period being the fasting season when vegetable consumption is usually higher. Based on our interviews, this seems to be linked to four factors:

    - There is a presumption among some urban residents that consuming raw vegetables increases the likelihood of contracting and spreading the virus, reducing demand for certain vegetables.

    - Some – especially larger and wealthier – traders are taking precautionary measures to avoid exposing themselves to the virus. These measures appear to have reduced their vegetable trading activity.

    - The travel bans have reduced the volume and frequency of trucks coming to Addis Ababa.

- Restaurants and other eateries have experienced a slowdown in business, and are opting to reduce vegetable purchases, among other responses.

2. **Urban retail prices are not significantly affected so far.** Both supply and demand appear to be impacted simultaneously: the lower urban demand that typically leads to a reduction in vegetable prices is balanced by a declining vegetable supply to Addis Ababa.

Effects on farmers

1. **Producer prices for vegetables are on the decline.** Fewer traders are traveling to rural areas because of the travel ban, the social distancing policy, and fear of infection. Combined with reduced urban demand and oversupply, producer prices are rapidly declining. For example, a quintal (100 kg) of head cabbage that sold for about 300 birr ($9) about two weeks earlier sold for only 100 birr ($3) at the end of March. Similarly, onions that sold for 15–17 birr ($0.50) per kg about two weeks earlier were selling for about 9–10 birr ($0.30) per kg at the end of March, a 40% decline. (A few key informants in these rural areas linked the price declines to a seasonal pattern, however.)

2. **Farm losses seem to be increasing.** A number of farmers we contacted indicated that they had to leave some vegetables in the field to rot due to the lack of buyers.

3. **There is a shortage of farm inputs and their prices are increasing.** Prices of important inputs crucial to vegetable production – including fungicides, insecticides, herbicides, fertilizers, and improved seeds – are increasing due to shortages. These seem to be linked to land border closings, which have blocked (sometimes illegal) imports from neighboring countries, and to reduced imports from China.

4. **Labor is becoming scarce.** Vegetable production is labor intensive and the Central Rift Valley, where most commercial production occurs, usually attracts a large number of daily laborers from across southern Ethiopia. These laborers often gather in set locations in rural towns to be picked up by vegetable producers. However, in response to restrictions on travel and gatherings (informally imposed by the regional police), these workers are increasingly returning to their home areas.

Conclusions and implications

The COVID-19 pandemic is beginning to disrupt food value chains in Ethiopia and elsewhere, impacting the livelihoods of farmers and the diets of rural and urban households. These effects are likely to hit the poorest and most vulnerable farmers and consumers the hardest, but they are not yet well understood. More evidence is needed to guide the government and other organizations in devising responses. While not representative of the whole value chain, our interviews with vegetable production stakeholders have a number of potential policy implications.

First, urban demand for fruits and vegetables – high-value, nutritionally rich foods – is declining. It is possible that this is driven by misinformation regarding the risk of contracting COVID-19 from produce. If so, there is a need for widespread and effective information campaigns.

Second, trade is affected by travel bans, as well as reduced competition, because traders are less willing to travel to production areas. Making sure that travel bans do not negatively affect food trade is paramount. To reduce the need for travel, enhanced trading through smartphones, virtual purchasing, and e-payments could be considered so that only truck drivers who pick up loads would need to travel, and traders and brokers would not need to travel.

Third, farmers are apparently hit in two ways. Producer prices are lower, and input prices are up or inputs are not available. Farmers will thus have less incentive to produce these crops, likely leading to lower yields and production in the near future. To avoid further disruptions to the food supply, ensuring the availability of agricultural inputs to farmers at low prices and assuring incentives for production should be a priority for the government in the next few months.

Ethiopia Strategy Support Program senior research fellow Alemayehu Seyoum Taffesse and IFPRI senior technical and policy advisor Anne Bossuyt also contributed to this post.

Originally published April 13, 2020.

19. Chinese livestock farms struggle under COVID-19 restrictions

Xiaobo Zhang

After the COVID-19 outbreak began in December in Hubei Province, many Chinese villages were locked down to control the spread of the disease. As the epidemic has eased, China has only begun to lift some restrictions. The lockdowns have had a significant – and still not well-understood – impact on the agriculture sector. The effective supply of agricultural products forms the foundation for a stable, functioning economy and safeguards people's livelihoods. Thus, keeping agricultural enterprises running is an indispensable economic component in the ongoing battle against the epidemic – yet discussions of the outbreak have thus far devoted very little attention to the challenges they face.

There are two key problems now.

First, livestock farmers face severe pressure from supply and market disruptions, since animals need to eat every day and production cycles are short – daily for dairy, six weeks for chickens, and three months for pigs. In addition, the pork industry is still reeling from the 2019 outbreak of African swine fever that reduced the country's pig herd by more than 40% and drove up prices.

Second, the arrival of the spring plowing season is putting crop farmers in a bind. They urgently need to return to work. But the outbreak and ongoing control measures present many challenges. While manufacturing and service enterprises can flexibly adjust their production schedules to mitigate losses arising from the epidemic, the agriculture sector waits for no one. The normal phase of spring plowing includes the provision of labor, seed, fertilizer, pesticide, and agricultural machinery, all within a set timeframe. Once smallholder farms miss the necessary services, such as plowing and pollination, during the critical farming season, their income for the entire year will fall.

To understand the operational situation and demands of small, medium, and micro enterprises impacted by the epidemic, the Enterprise Survey for Innovation and Entrepreneurship in China (ESIEC) project team conducted telephone and online follow-up interviews in February with enterprises surveyed over the past three years, which include some in the agriculture sector. The survey probed their work resumption and production situation, the main difficulties they face, their efforts to adapt, demands for appropriate policies, and other issues.

Most Chinese farms are household smallholders and not registered as enterprises. Therefore, our survey does not capture the direct impact on crop farms. But other types of agribusinesses, such as mechanization services, pollination services, fertilizer dealers, and livestock farms, are covered in our sample and provide a broad picture of the impacts. As of February 10, only 24.6% of agricultural businesses had resumed production. Since these businesses provide key inputs or services, interruptions in their services may negatively impact agricultural production.

In our analysis, we found that the main issue agricultural enterprises face is disruption of logistics, especially shortages of raw materials and delivery problems. The stress is particularly acute for livestock farmers: 38.5% of them list "logistics disruption" as the biggest challenge, compared to 35.6% of all agricultural enterprises, 19.7% of non-agricultural enterprises, and 18.9% of the service sector.

Shortages of raw materials – in particular an inadequate supply of feed to livestock farmers – are the main result of these disruptions. While about 60% of the agricultural enterprises surveyed have encountered such shortages, they are most severe in the livestock farming sector – where feed shortages mean that animals and poultry may starve to death. With the preexisting problems and lingering high prices from the swine fever outbreak, the industry faces a crisis that could lead to more price spikes.

Overall, respondents' two most common complaints were that feed could not be delivered to the farms, and that trucks could not enter villages to collect their products.

The results can be dire. On February 13, for instance, a beekeeper in Sichuan Province committed suicide after his bees starved to death because his truck of beehives was not allowed to travel across regions to provide pollination services as scheduled. The lack of pollination services may lead to lower yields for many crops.

What policies are needed to address these problems? More than other industries, agricultural enterprises and particularly those in the livestock farming sector say they prefer rent reductions, financing support, and especially *force majeure* certification.

Our survey makes it clear that, apart from the above options, a simpler and more direct demand often expressed by the entrepreneurs is that the lockdown be ended. The Chinese government has already rolled out a series of pertinent measures, including opening a "green channel" for feed, in order to effectively stabilize agricultural production. However, this has not yet been implemented in all areas. While all relief efforts are important, it is even more urgent that these measures be implemented at the grassroots level. While maintaining effective control of the epidemic, it should be of the greatest importance to encourage enterprises in rural areas to return to work.

Given that other countries and regions around the world have also adopted lockdown policies, the survey results suggest that as spring arrives, agricultural enterprises in many places face serious logistics problems, and that livestock farming also faces challenges similar to China's – problems that may require government intervention to avert shortages or price spikes.

The chapter is based on a joint research paper with Zijun Cheng. We are grateful for the support from a special grant for studying the impact of COVID-19 from Peking University, a research grant from China Natural Science Foundation (#71874008), and the CGIAR Research Program on Policies, Institutions, and Markets (PIM), led by IFPRI. I would also like to thank the Enterprise Survey for Innovation and Entrepreneurship in China (ESIEC) Project Alliance (formed by Peking University, Central University of Finance and Economics, Harbin Institute of Technology at Shenzhen, Guangdong University of Foreign Studies, and Shanghai University of International Business and Economics) for conducting the survey and allowing us to use the data. We have received helpful comments from the editors.

Originally published March 26, 2020.

GENDER

20. Why gender matters in COVID-19 responses – now and in the future

Agnes Quisumbing, Neha Kumar, Ruth Meinzen-Dick, and Claudia Ringler

To contain the spread of COVID-19, health ministries and the World Health Organization (WHO) are advising everyone to keep up to date on latest developments, wash hands frequently, stay at home, and practice physical distancing when outside the home.[1] These recommendations are inconveniences for most people in Europe or the United States, but for many in developing countries, even these basic precautions will be difficult to implement.

Here are some ways these public health recommendations affect women and men differently in developing countries, particularly in rural areas – and some ideas for how to address the disparities.

Stay informed

WHO recommends that everyone keep up to date on the latest information on COVID-19. This is a particular challenge for rural women, who have lower literacy and numeracy rates and less access to modern information and communications technologies. Mobile phones are seemingly ubiquitous, yet out of more than 2 billion people in low- and middle-income countries, only 82% of women own one – meaning 393 million are excluded, mostly in rural South Asia and Africa. Even women with access may not have their own phones, and tend to use a smaller range of services.

Key barriers for women include affordability; literacy and skills needed to use the device; safety and security (including personal safety) when using the device; and lack of family approval. The gender gap tends to be particularly high in rural areas. To address these disparities, IFPRI, together with partners in Kenya (Groots Kenya), India (Self-Employed Women's Association-SEWA), and Uganda's extension service, is testing alternative ways to reach women farmers with information, including WhatsApp, posters, and videos. Some countries and organizations are providing free cellphones or airtime to women to support them during the crisis.

Hand-washing

Frequent hand-washing with soap is a key measure in the fight against COVID-19 – but out of reach for many households. In 2017, 3 billion people still lacked basic hand-washing facilities at home: 1.6 billion had limited facilities lacking soap or water and 1.4 billion had no facility at all.

1 Other recommendations include respiratory hygiene and to avoid touching eyes and mouth as well as to seek medical help early if needed. The reference was last accessed June 4, 2020, and it is frequently updated.

Unsurprisingly, this deprivation falls mostly on the poor; nearly three-quarters of the population of least developed countries lacked hand-washing facilities with soap and water.

The task of procuring water for hand-washing and other domestic uses falls disproportionately on women and girls. Strict lockdown rules in many countries, including curfews and limits on congregating at common water distribution points, further compound these problems.

There are, however, promising interventions. In Bangladesh and Uganda, for example, the "tippy tap" – a simple, low-water-usage device – has been promoted to improve hand-washing, and combined with behavior change communication (BCC) targeted to women. IFPRI, under the CGIAR Research Program on Water, Land and Ecosystems, has supported cost-effective social learning interventions to change sanitation behavior, with lessons for COVID-19.

Stay at home

"Stay-at-home" recommendations and the strict lockdowns in many countries have left both men and women jobless. Many migrant workers also lost their jobs and had to return to rural homes. In many contexts (for example in Middle Eastern and North African countries, India, Nepal, and Tanzania), women whose husbands migrate gain autonomy in decision-making, which is often cherished despite the increase in responsibilities. As male migrants return home, women suddenly lose this autonomy and their role as the de facto household head.

On top of the financial stress to individuals and families, confinement can lead to mental stress. For men, who are typically seen as and consider themselves to be their families' breadwinners, loss of employment and income may result in mental health problems and/or domestic violence as an outlet. Men may also lose contact with their peers, exacerbating stress.

For women and children, quarantine conditions thus increase tension and exposure to potential perpetrators. Overburdened health services – often the first point of contact for women experiencing domestic violence – may be unable to respond. A comprehensive review by Peterman et al. (2020) identifies potential direct and indirect pathways between pandemics and violence against women and girls – including effects on economic insecurity and poverty-related stress, increased exposure to exploitative relationships as household structure and composition change, and the inability of women to temporarily escape abusive partners.

Stay-at-home orders also make it difficult for many women to procure food for cooking, one of their key responsibilities directly affected by COVID-19. Women may have to prioritize the limited amount of time permitted outside the home among a number of tasks, choosing between foregoing procuring safe water or food or fuel for their children and families. And food insecurity may affect women more than men, as seen in previous work on the food price crisis of 2007–2008.

Physical distancing

Yet we should not underestimate the resilience of women's groups. PRADAN, one of India's largest NGOs, is using these groups as a platform for community kitchens, providing meals to those in need (especially migrants returning from urban centers). SEWA is developing risk communication and community engagement plans using grassroots leaders and WhatsApp to educate members about protecting their families' health. Thus, while conditions are more difficult, existing women's collectives are proving a valuable asset in the pandemic response, and may take on potential new roles such as testing and contact tracing. Gender-sensitive programming could also look into supporting men in their care-giving roles as well as providing psychological support.

Other consequences of the pandemic: Illness, death, and loss of schooling

Our previous work in Bangladesh and Uganda shows that shocks like illness and death affect men and women differently. The burden of caring for the sick falls disproportionately on women's shoulders, so in the short term, their exposure to sick individuals may increase their risk of contracting the virus. Moreover, women's assets may be sold first to cope with illness; in the longer term, such losses may leave them more vulnerable to future shocks. Emerging evidence seems to indicate that men are dying of COVID-19 at higher rates than women, possibly due to a combination of biological and social factors. The death of an income earner may severely affect women, depending on inheritance patterns and practices upon marital dissolution (whether through death or divorce).

As their small businesses collapse and their informal work arrangements are cancelled, women will lose financial independence, affecting their empowerment in the short term, with potential longer-term impacts on children's schooling (particularly for girls). This, in turn, could affect female labor-force participation in the next generation.

Strengthening women's assets should be a key priority in pandemic response and recovery. Because women's assets are often the first sold in economic crises, protecting them to the extent feasible and rebuilding them following COVID-19 will be crucial. Such efforts also support women's empowerment. Rebuilding the social capital embedded in women's groups may also empower women to be more aware of and to avail themselves of public services, and to provide the leadership their communities need.

While WHO's COVID-19 guidelines are essential for everyone's health, it is clear that women face challenges in implementing them that are quite different from those faced by men. Women need support from governments and international health and women's organizations – now and in the future – to ensure that the pandemic does not wipe out decades of gains in women's empowerment and family well-being.

The CGIAR Research Program on Policies, Institutions, and Markets (PIM) provided support to IFPRI research cited in this post.

Originally published April 22, 2020.

21. Why gender-sensitive social protection is critical to the COVID-19 response in low- and middle-income countries

Melissa Hidrobo, Neha Kumar, Tia Palermo, Amber Peterman, and Shalini Roy

Many governments are using social protection programs to respond to the economic crisis and health risk induced by COVID-19. As of April 17, 133 countries had adapted or introduced 564 social protection initiatives, according to the World Bank. With the focus on rapid assistance, gender considerations have understandably not been at the forefront of these efforts. A rapid assessment of initial COVID-19 social protection responses indicates that only 11% show some (albeit limited) gender-sensitivity.

This is unsurprising – most existing social protection programs in low- and middle-income countries (LMICs) are either gender-blind or neutral at best – but it is worrying. The COVID-19 crisis has the potential to widen gender inequalities, including those related to loss of livelihoods, reproductive health risks, disproportionate burden of care, and violence against women and children. Social protection that does not take gender into account can reinforce these inequalities.

General guidelines for COVID-19 social protection responses are available, but how can governments address gender inequalities? Designing gender-sensitive programming is not always straightforward, but evidence suggests simple design and implementation adaptations can make programming more gender-sensitive. While there is no one-size-fits-all approach, in a new brief summarized below, we provide key lessons, considerations, and guidance across five areas.

1. Adapting existing schemes and choosing the forms of social protection

Adapting existing schemes to be contagion-safe is a likely first step for governments, and these adaptations can have gender implications. Relaxing existing conditions (for example, those tied to work, health, or schooling) can simultaneously reduce viral spread and benefit women who are often responsible for fulfilling conditions, may be mobility-constrained, and may have fewer social or information networks.

Expanding access to healthcare via fee waivers or providing automatic health insurance enrollment can support women in continuing to seek care for critical, routine maternal and child health and reproductive health services. Cash benefits (via e-payments) are widely recommended; cash can also improve household economic security and emotional well-being, which directly benefit women and can contribute to reducing intimate partner violence. However, the feasibility of safely providing additional in-kind transfers (including food or soap) should be considered as well, as women and children are often the first to reduce food consumption in response to food insecurity, and women may be responsible for daily shopping, exposing them to potential infection. In-kind transfers should be considered where mobility is restricted, markets are limited, food prices spike, or COVID-19 restrictions induce supply chain closures.

When social distancing restrictions are relaxed, implementers of public works programs should ensure dignified work with fair wages where women can safely participate, with exemptions for lactating and pregnant women. When schools reopen, implementers should pay particular attention to re-enrollment of adolescent girls and relax economic constraints with appropriate policy instruments.

2. Targeting

How to target households and individuals are critical considerations. Retaining the original individual-level targeting of many existing programs may be most straightforward; however, such targeting can exclude vulnerable populations. For example, unemployment insurance typically does not cover informal workers, including the majority of women who primarily work in the informal economy. Providing universal household-level transfers can reach more vulnerable people, but who the household's "named recipient" is may also have gender implications. Although broader evidence is mixed, a few studies from LMICs indicate that naming female recipients may improve women's empowerment. We believe the evidence supports considering women as named recipients – while recognizing that particularly acute periods of the crisis (such as lockdowns) may intensify household tensions.

Therefore, in settings where existing analysis shows the feasibility and acceptability of targeting women, we see gains in continuing during the COVID-19 crisis. But in settings where targeting women was previously deemed infeasible, we do not recommend starting during the crisis and explicitly challenging norms during a time when tensions are high. Nonetheless, even in the latter case, minor tweaks in operationalizing targeting – including authorizing multiple household members to make transactions, ensuring information reaches both men and women, and providing messaging that benefits are for the entire family – could contribute to greater gender equity.

3. Benefit levels and frequency

Benefits in response to COVID-19 should be quick and lumpy, ensuring sufficient support before supply chains are overwhelmed – and to avoid health risks from more frequent payment distributions and contact. While qualitative studies indicate that women may be able to retain control of smaller transfers, large randomized studies suggest larger cash transfer values result in higher benefits for households and women specifically. In addition, no studies we are aware of show that larger transfers to women induce adverse effects.

Therefore, we believe programs should provide sufficiently high benefit levels to cover the duration of the COVID-19 economic crisis, understanding that programming during this time may be a full income replacement, rather than supplement. In addition, "top ups" should be considered for households caring for sick members or children to address disproportionate care burdens. Finally, it is important to consider that female-headed households are often smaller – and thus may appear better-off in a direct per capita poverty measure – yet may still be more disadvantaged for numerous reasons (for example, discrimination or access to services).

4. Delivery mechanisms and operation features

Programs generally employ the most logistically feasible delivery mechanisms and operational features in crisis conditions, but seemingly simple choices may have gender implications. Accessible grievance mechanisms should be set up, and implementation and management staffing should include women. Delivery mechanisms for benefits and information should be practical and accessible to both men and women.

While e-payments may not be an option for many settings, in the longer term, national programs should invest in these. Extending the network of e-payments may increase financial inclusion, including among women, who have lower inclusion rates. Responses should consider that in many settings women are less likely to have access to mobile phones; existing programs have sometimes provided them for this reason. While mobile phones are a promising platform for providing information, it is important to keep in mind that improving access alone may not be sufficient; women also have lower literacy, lower ability to pay for services, and multiple constraints on their time. Thus, mobile phone-based platforms should be complemented by other platforms such as internet, television, and radio; and when possible in mobile platforms, voice messages or speaking directly to an expert are preferred to text messaging. Women's groups or other peer support groups may be leveraged as networks for more efficient communication and delivery of essential services.

5. Complementary programming

Complementary programming remains relevant for women during COVID-19, especially on topics related to food and nutrition, including ways to access or grow nutritious foods when markets and supply chains are down; water and sanitation, as information about hygiene and social distancing is critical for reducing COVID-19 spread; maternal health including antenatal care, as travel may be restricted and health centers overburdened and a potential infection risk; sexual and reproductive health, including family planning and menstrual hygiene management; parenting and learning for children as many schools are closed; mental health for both men and women, given that many may experience depression related to isolation or loss of livelihoods; and access to referrals for violence-related services.

All of these comprehensive services will rarely be available, particularly during the pandemic, but social protection platforms can at a minimum explore integrating light-touch information campaigns with delivery taking into account the gender considerations outlined above and linkages to services.

Concluding thoughts

The COVID-19 pandemic presents an opportunity to address existing gender inequalities through social protection. Program designs should be adjusted to account for gender, in a manner informed by existing analysis, while taking a long-term approach. Related issues of political economy, coordination, and financing that have gender considerations should be explored in future guidance. Because these are complex issues and unintended consequences of programming are possible, more research is needed on intersections of social protection, gender, and pandemics, where ethically feasible. At a minimum, monitoring statistics should be sex- and age-disaggregated and, where possible, data should be collected to ensure risks to beneficiaries do not increase. Taken together, these policy adjustments and new evidence can lay the groundwork for more gender-sensitive social protection systems in LMICs both during the crisis and beyond.

This work was undertaken in collaboration with the Transfer Project and as part of the CGIAR Research Program on Policies, Institutions, and Markets (PIM) led by IFPRI.

Originally published April 28, 2020.

RISE
FALL

POLICY RESPONSES

22. Fiscal and monetary responses to the COVID-19 pandemic: Some thoughts for developing countries and the international community

Eugenio Díaz-Bonilla

Most developed countries have implemented massive economic responses to the COVID-19 pandemic, ramping up spending and using monetary policy to cushion the blow of lockdowns and other measures that have shut down businesses and left huge numbers unemployed. But for developing countries, which are now starting to respond to the crisis more aggressively, such options may be more limited. Here I discuss some ideas for how these countries may address the economic fallout, and how international organizations can help.

Addressing the crisis in developing countries

Developing countries should start by implementing a national response plan focusing on these four interrelated spheres: health; the supply and demand of essential goods and services; the domestic financial circuit in local currency; and the foreign currency market, linked to international trade and external debt. Such a plan requires a centralized crisis-management office led by the president, prime minister, or equivalent, with participation of the relevant public and private sector representatives. This is easier said than done, but it is the only way to avoid uncoordinated actions and working at cross purposes.

Here I will focus on the latter three spheres, acknowledging that there are interactions with health measures, as well as short-term trade-offs between health controls and economic activity.

Supply and demand of essential goods and services

Governments must address basic supply and demand issues to prevent shortages, price spikes, and economic disruptions in the short term. It is essential to ensure the production and distribution of food and medicines, which in turn requires keeping transportation and basic public services (water, energy, and communications) up and running. The crisis-management office must establish committees with the private sector and operators in key areas to monitor daily the flow of crucial goods and services, and the health of workers and critical personnel. Bottlenecks, as well as hoarding and unfair trade practices, must be monitored and energetically addressed.

Regarding demand, governments must enact initiatives to support employment and income, including expanding safety nets with a food component. Central banks can play a key role, pursuing unconventional monetary policies that establish various channels to inject liquidity into the economy (I discussed several options in Díaz-Bonilla 2016 and 2018, for a different context). Of course, developing countries with a very strong fiscal position pre-COVID-19 may be able to borrow in domestic and foreign currency, without immediately resorting to the approaches discussed. What follows assumes that such is not the case in most developing countries.

The domestic financial circuit in local currency

Supporting supply and demand for basic goods and services will require an expansion of the money supply (see Díaz-Bonilla 2015). Central banks must dust off the instruments they used when they were called "developmental central banks." This means taking steps that exceed recent monetary interventions, and which may raise objections.

The recent expansion of the "quantitative easing" approach to increase money supply has nonetheless remained more limited in scope and impact than past options, in developed as well as in developing countries (see a general discussion in Epstein 2005; the specific case of the US Federal Reserve is in Fettig 2008). In fact, the process of creating "modern" central banks has mostly involved restricting monetary instruments, mainly because of concerns about their past use (and abuse) leading to high inflation in many developing countries; because they may amount to picking winners and losers; and, perhaps, because of distributive effects. The latter two effects are unavoidable with any monetary mechanism, however indirect (see Coibion et al. 2012).

Nevertheless, central banks must expand their options for lending to the private sector and to the government.

Part of the private sector support can be offered through rediscount credit lines to banks so that they, in turn, may maintain soft lines of credit for the working capital of companies, especially small and medium-sized enterprises (SMEs), including small and family farms. Those soft rediscount lines (or even outright grants, using a non-bank channel) should require businesses to keep employees on the payroll. In particular, these lines of credit could be crucial to support operators in food systems (especially family farmers), the health sector, and other crucial activities.

Central banks can also finance the public sector directly, with the objective of expanding food programs and safety nets (including considering some form of universal income), supporting the operation of the health system, financing other basic services, and investing in public works. These initiatives will definitely expand the money supply. That in turn requires eliminating or reducing other sources of money creation, on the one hand, and trying to align the supply and demand for local currency, on the other. The latter is necessary to avoid a spike in inflation and/or a currency run (discussed below).

If the supply of basic goods and services is ensured, as discussed above, inflation risks will be reduced. Also, the economic deceleration or recession from the pandemic shock, and the increase in what economists call the need for "precautionary balances" or more savings by households and individuals, would work against an inflationary shock.

During this process, the banking system must be monitored continuously to ensure its proper functioning. There will be need of some flexibility for debtors, but also for the banks when they are evaluated and audited by the central bank or equivalent authority.

Supply and demand for foreign currency

To avoid a run on the domestic currency from the expanded money supply, governments will most likely have to establish controls on transactions in foreign currency. The government must be able to manage foreign reserves, calculating the cash flow needed to finance the imports of food, medicines, energy, and other basic materials for at least six months, while considering the net flows of external debt.

One crucial consideration: avoiding an overvalued official exchange rate. Many developing countries' exports will decline in price and quantity due to lower world demand, and remittances will also be affected. Addressing that shock requires maintaining an official exchange rate valuation that does not discourage exports needed to finance crucial imports and other external flows. Also, developing countries may reduce import taxes for critical goods (to alleviate inflationary pressures), while refraining from export bans on food and other basic products.

The role of international organizations

During the 2008-2009 crisis, the G20 suggested a variety of domestic and international measures to confront the global recession. Domestically, it called for strong monetary and fiscal stimulus. Internationally, it expanded the capital base and operations of the international financial organizations. Now that developed countries are taking stronger and more unconventional monetary approaches, the G20 and the United Nations (UN) should respond vigorously on behalf of developing countries.

International organizations must call for further rounds of "unconventional monetary policies" coordinated with fiscal stimulus in developing countries, as discussed above, allowing them the policy space to decide how to do this (in many countries, this would most likely mean separating the local and foreign currency markets). They should also encourage leaders in developing countries to establish a central crisis management office as outlined above.

Those domestic responses must be supported by international action from the UN and the G20:

- **Increases in capital at the International Monetary Fund (IMF) and multilateral development banks (MDBs)** (at least the same amount as in 2008-2009). This will take time to negotiate, so the MDBs can be asked to adjust their financial policies to be able to increase their loans/equity ratios (say, up to 5 or somewhat more). This will require establishing the regulatory mechanisms and dialogue with credit rating agencies to adjust their criteria for risk rating, so these changes do not lead to downgrades of ratings of the MDBs, which would constrain their lending capabilities just in the middle of the recession. The additional lending capacity in MDBs should focus mainly on financing the health budget of developing countries; strengthening safety nets; financing food supply and distribution; and financing working capital for SMEs.

- **An additional allocation of special drawing rights (SDRs)** to about double the amount of 2008–2009.

- **Establishing a mechanism of debt resolution for developing countries** focusing, at least, on the debt coming due in the next two years. At a minimum, it is necessary to ensure neutral capital flows with MDBs and bilateral financial agencies, along with the rollover of private debt coming due in that period. Many developing countries will need further support due to larger trade deficits and declining remittances.

- **Directing MDBs to enter into conversations with private sector banks and investors to establish different mechanisms of co-lending** (such as what are called A/B loans, selling part of the developmental portfolio, and similar options).

- **Directing the UN agencies and the MDBs to set up mechanisms to advise and support developing countries in their policy responses** within the health sphere and in the real-economy sphere, particularly food, medicines and health equipment, energy, and basic public services.

It would also help if expanded liquidity swaps (as done in 2008–2009) could be established by central banks across a larger number of developed and developing countries. Initial movements in that direction are underway.

Conclusion

These historically unprecedented times require unconventional responses. Yes, there are several examples of countries that in the past have abused "unconventional monetary approaches," leading to high bouts of inflation, strong devaluations, balance of payment crises, and corruption. Yet, with prudence, these approaches should now be used to finance specific public expenditures, such as cash transfers and safety nets for the poor and vulnerable, and certain public investments, and to keep firms operating. In any case, money always enters into the economy through specific actors (at present, mainly the owners of the assets being bought by the central banks), and not by equally endowing each citizen with the same amount of currency (as in Milton Friedman's parable of "helicopter money"). A universal income would do the latter, and some of the recent rescue packages in developed countries moved in that direction. The methods suggested here would ultimately make the channels through which an expanded money supply gets into the economy more democratic.

The international community must also step up its response as outlined above.

The situation is very complicated, and the world will not be the same after this crisis. It is in our hands to limit the global damage and to establish the foundations for a strong rebound afterward.

Some of these ideas appeared in an article in Clarín (Buenos Aires).

Originally published April 5, 2020.

23. Social safety nets are crucial to the COVID-19 response: Some lessons to boost their effectiveness

Daniel Gilligan

The twin health and economic shocks of the COVID-19 pandemic are staggering in their breadth and scale. While the disease arrived later and has spread more slowly in many low- and middle-income countries, COVID-19 is threatening the lives and long-term livelihoods of millions of poor people, and could push an additional 140 million into extreme poverty.

We know from past crises that while an economic shock's impacts may vary across the income distribution, the poor face a multitude of vulnerabilities to the pandemic's effects. In many places, the poor are more likely to have underlying or untreated health conditions (such as hypertension, diabetes, and vitamin D deficiency) which raises the risk of serious illness if they are exposed. Poor households are also more likely to cope with income loss by selling productive assets or undertaking work that is either inherently riskier (for example, construction or sex work) or increases their risk of coronavirus exposure. Their children may also be less likely to return to schools upon reopening, permanently reducing their earning potential. These factors leave the poor even more vulnerable to additional impending shocks, like desert locusts in the Horn of Africa or cyclones in South Asia.

Targeted social safety nets for the poor are central to the effort to stifle these negative impacts and protect the substantial gains made globally in the fight against poverty, food insecurity, and malnutrition in this century. Why are social safety nets important to the pandemic response, what challenges do they face, and what lessons can we glean from past research into social protection programs to craft effective responses over the long term?

Social safety nets have played a major role in the response to the COVID-19 pandemic in the last three months. According to an effort by Ugo Gentilini of the World Bank to track social protection responses during the crisis, 190 countries have implemented, adapted, or planned over 900 social protection measures during the crisis, often in the form of cash transfers. This response has included an expansion in the number of social protection beneficiaries by roughly 15% in South Asia and East Asia and the Pacific, but by only 2% in the Africa region. Several countries have offered a temporary sharp increase in the benefit amount for current beneficiaries.

Why social safety nets during a health crisis?

Three factors explain why social safety nets have been central to the COVID-19 response. First, two decades of extensive, rigorous research on social protection programs has documented their effectiveness at protecting food security, assets, and human capital, including in a crisis. Safety nets can also improve health, including for newborns, through improved nutrition when combined with complementary nutrition programs, a high policy priority during the pandemic.

Second, large-scale transfers help to replace lost income for credit-constrained poor households and counter the economic drag of the pandemic by providing a fiscal stimulus that, under certain conditions, may generate positive multiplier effects during the recovery. A temporary increase in cash transfers during the pandemic can also make moral and political sense, protecting those most in need and building trust in government.

Third, the infrastructure of a social safety net was already in place, to varying degrees, in most countries. Over the last two decades, social protection, and particularly social assistance, has grown in popularity as a leading response to poverty. Many poor countries developed targeted in-kind and cash transfer programs, expanding their roles and improving their efficiency. Many safety net programs are designed to be "shock responsive": During a 2011 drought, Ethiopia increased benefits for 85% of the 7.6 million beneficiaries in the Productive Safety Net Program (PSNP) and provided temporary benefits to an additional 3.1 million people. As the COVID-19 pandemic hit earlier this year, many countries had the payment mechanisms, poverty registries, and local selection committees in place to quickly expand benefits and renew program targeting.

Challenges ahead

For countries that have had initial successes using social safety nets to respond to the pandemic, significant challenges remain. For most, the fiscal cost is the greatest of these; continuing transfers beyond a few months will be difficult. Yet the pandemic and its economic consequences may drag on, or return later this year after a failed reopening. Social distancing measures pose obstacles to targeting new beneficiaries, as this typically involves in-person interviews or local committees meeting to screen applicants. Delivering cash or in-kind transfers can also increase the risk of coronavirus transmission, and most programs still deliver benefits in ways requiring person-to-person contact.

How should social safety nets be designed for the pandemic?

Evidence from past research on the design, implementation, and effectiveness of social assistance programs during an economic crisis provides guidance on handling some of these challenges:

1. **When safety net programs are disrupted, maintain transfers by adopting alternative delivery methods.** At the start of the pandemic, many school feeding programs ceased operating as countries closed schools. These are one of the most popular forms of social assistance, reaching more than 270 million children in 89 countries in one World Bank estimate. In India, many school systems replaced on-site meals with rations delivered to homes or available for pick-up at schools. As they reopen, schools should consider maintaining take-home rations initially, to reduce the risk of coronavirus transmission when students eat together; in Uganda, fortified take-home rations were as effective as equivalent on-site school meals for improving school participation and attainment, and for reducing anemia prevalence for adolescent girls.

2. **Eliminate conditions on assistance temporarily.** Reassess their importance and effectiveness during the recovery. Cash transfers are often conditioned on child school participation or adult work requirements. These should be suspended during the pandemic. Unconditional cash transfers can also improve school attendance. In Ethiopia, PSNP work requirements were temporarily suspended for beneficiaries because it was no longer possible to bring work teams together safely. Payments to public works beneficiaries continued, and were even accelerated in case it became impossible to make payments for some time. When work can resume, programs should identify work activities that bolster the COVID-19 response, such as building or renovating health clinics or supplying hand-washing stations.

3. **Strengthen and expand targeting.** Safety nets often exclude a high proportion of the poorest households and fail to reach the most vulnerable groups. They should expand coverage in poor areas and make efforts to target migrants, orphans, and the urban unemployed, who face substantial livelihood risks and have limited social support. Social assistance programs should also be used as platforms to identify the newly poor. For example, community health systems and local government can be used to target new beneficiaries. In areas that are particularly hard-hit with a collapse in employment, programs should consider providing universal transfers temporarily.

4. **The form of assistance matters, but distributing it quickly is the priority.** Food, vouchers, or cash assistance can all increase household consumption and improve calorie intakes, though vouchers may perform better than food transfers or cash at improving measures of dietary diversity and thus quality. However, setting up a voucher system involves coordination with food retailers and thus may be challenging to establish during the pandemic. Cash transfers have many of the advantages of voucher programs – they work well where people have access to markets and as long as prices do not escalate – and they are usually cheaper to deliver. Still, all of these transfer modalities can be beneficial, so following a "no regrets policy" is best: Prioritize speed in scaling up transfers with whatever method works the fastest.

5. **Respond to the crisis with the future in mind.** Many acknowledge the need to "build back better" following the pandemic to improve safety net systems. Here are some priorities:

   - *Strengthen mental health services for beneficiaries.* Anxiety and depression increase with poverty and, in women, are associated with worse development outcomes for their children. The pandemic is an opportunity to bring mental health out of the shadows. Programs should build a cadre of community health workers to provide mental health services and screening.

   - *Design new programs or features to be gender-sensitive.* Most social assistance programs are not designed with the specific needs of women beneficiaries in mind, a wasted opportunity that can widen gender inequality. As my colleagues Melissa Hidrobo, Neha Kumar, Tia Palermo, Amber Peterman, and Shalini Roy emphasize, social protection programs designed to be gender-sensitive during the pandemic have the potential to protect women's livelihoods and assets, reduce unequal burdens of care, improve women's empowerment, and reduce intimate partner violence.

   - *Strengthen nutrition-sensitive social assistance.* An ever-growing body of evidence shows that critical investments in health and nutrition for the poor can be better achieved by combining cash transfers with information campaigns and access to services. These investments in child human capital will have large economic returns as these individuals enter the labor market and begin their own families.

- *Invest more in mobile payments.* Contactless mobile payments have a clear advantage during the pandemic. Though the poor continue to have less access to mobile phones, it is important to provide transfers through mobile phones when possible – or better yet, to steeply subsidize mobile phone ownership for the poor. Many will benefit from the new access to social networks and information. Even where literacy is low, household heads often have a family member who can support the phone use.

6. **Strengthen fiscal support for the social assistance response.** The global slowdown in economic activity and lockdowns have ravaged economies of low- and middle-income countries at a time when many face a debt crisis. With public health costs growing enormously, governments will need aid to finance a robust social safety net response.

As they try to contain the pandemic, countries must also confront a rise in extreme poverty and the suffering that goes along with it. But they have many tools to combat this problem. Extensive research on social assistance programs provides a firm foundation for approaches to strengthen safety net designs and adapt responses to fit changing circumstances. These actions can improve social safety nets and help to counter some of the worst effects of the COVID-19 pandemic.

Originally published June 18, 2020.

24. How India's food-based safety net is responding to the COVID-19 lockdown

Devesh Roy, Ruchira Boss, and Mamata Pradhan

India's huge population, its density, and very large numbers of poor present an extraordinary challenge for the country's COVID-19 response, and the Indian government imposed the largest lockdown in history: 1.3 billion people ordered to shelter in place for 21 days as a part of "lockdown 1.0" which began March 25. Even China, where the disease originated, ordered a total lockdown in just one area, Hubei Province (while imposing other restrictions throughout the country). Implementing a lockdown in a country of India's size has been socially, economically, institutionally, and politically very demanding and disproportionately affects the poor, daily wage earners, and other marginalized groups.

Thus COVID-19 exposes a harsh reality: an inadequate and uneven safety net may leave many from these economically vulnerable groups without access to food and other services. This struggle is particularly acute for large numbers of informal sector workers – including self-employed, subcontracted laborers, small farmers, and landless workers. India's informal sector employs 303 million; the workforces of Uttar Pradesh and Bihar states are more than 80% informal, while even in advanced states like Maharashtra, that share is 70%. COVID-19 may push this group and their families into transient poverty.

The nature of the COVID-19 pandemic is unique. Shutting down many business operations, which leaves people without work, is an integral part of efforts to "flatten the curve" of disease progression. Laid-off workers, particularly daily wage workers who are largely seasonal migrants, will struggle to find employment even as the lockdown eases. Some 9 million workers are estimated to move annually, though their total number was as high as 139 million in 2011. Coronavirus-related layoffs will disproportionately hit service workers in low-paying jobs as restaurants, malls, cafes, and shops shut their doors indefinitely.

For informal sector workers and rural poor, missing even a day's earnings can make it difficult to buy basic food items, and joblessness extended over several days can mean economic ruin. As India witnesses large-scale reverse migration, with desperate migrants leaving cities amid lockdown and walking hundreds of miles toward their home villages, the prospect of economic devastation and a growing population of rural poor – internal COVID-19 refugees – looms large.

Yet compared to those of other countries at a similar income level, India's social safety net is extensive. An elaborate array of programs exists to assist the poor, including the world's largest food-based social program, the Public Distribution System (PDS), covering 800 million people. To respond quickly, India is utilizing these existing schemes and reshaping them to address the unique challenges of COVID-19.

The food-based safety net and COVID-19

On March 26, the government announced a $22.6 billion relief package with a major food component. PDS has played a key role, providing 5 kg of either rice or wheat and 1 kg of preferred pulses per month free, offered in two installments. (This is in addition to the preexisting entitlement of 5 kg of low-cost wheat/rice per person per month.) The relief package cereal allotment should meet most families' cereal requirements, but the pulses allocation is likely inadequate, given that per month consumption is 4 to 5 kg. Several states, meanwhile, have announced their own relief packages (Table 1). Supplying these relief efforts should not be a problem. Rice, wheat, and pulses stocks are adequate to feed the country for now, and the harvest of rabi crops is around the corner. This emergency food support is happening at a propitious time.

Implementation challenges

Implementing the relief package is fraught with challenges that must be addressed, or the entire effort could be undermined:

- PDS coverage in urban areas is low (about 50%), thus leaving out many urban poor. In response to the pandemic, responsible agencies should quickly expand the list of eligible households. If necessary, the broader coverage could be rolled back after COVID-19 subsides. In the initial phase of the lockdown, only the Delhi government had announced that people without ration cards could also get rations. The central government made a similar announcement of free distribution of 5 kg each of wheat and rice and 1 kg of pulses per family to 80 million migrant workers without ration cards almost a month and a half following the first lockdown. Bihar announced use of direct cash transfers, depositing funds to the bank accounts of ration-card holders.

- With commodity prices expected to rise, and the small amount of pulses in the relief package, ensuring access to adequate diets is problematic. Indeed, some states are offering a food kit via PDS (see Table 1 for food kit intervention).

- Given the stresses of the emergency, there is a high likelihood the program will have both exclusion and inclusion errors.

- PDS ration cards are neither portable across locations nor can rations be divided to allow family members to pick up portions at different locations; this makes them potentially useless for seasonal migrant laborers. The government's new One Nation, One Ration Card (ONORC) program will be rolled out in all states and Union Territories only by March 2021. Even if implemented then, the program has no provision for divisibility.

- Food quality, not just quantity, must be maintained. India's supply chain must gear up to deal with the transportation, storage, and distribution of large volumes of food in short timeframes during the lockdown to avoid spoilage and contamination.

- The extra grains pumped into the system are likely to depress prices; in the long run this may affect small farmers and small businesses – another impact on those most vulnerable to COVID-19 restrictions.

TABLE 1 State governments' food-based relief measures during COVID-19 lockdown

STATES	TOTAL BENEFICIARIES	SUBSIDY/ FREE	QUANTITY (PER MONTH)	DURATION (MONTHS)	INSTALL-MENTS	ADDITIONAL INFORMATION
Kerala	NFSA ration-card holders/people under quarantine/passengers stranded in the state	Free (canteen meals subsidized)	BPL families to get 30 kg of rice along with the kit/ those entitled to 2 kg of rice will get free 15 kg rice/hh & food kit (sunflower oil - 1 kg; coconut oil - 1/2 kg; salt - 1 kg; wheat flour - 2kg; rava - 1 kg; green gram - 1 kg; black chana - 1 kg; tur dal - 1/4 kg; mustard - 100 gm; fenugreek - 100 gm; coriander - 100 gm; urad dal - 1 kg; chilli powder - 100 gm; sugar - 1 kg; tea - 250 gm)	1		1,000 food canteens for subsidized meals at Rs. 20. Open community kitchen to be formed at the panchayat level will operate across the state: a telephone number will be given for people to reach out and ask for food. Passengers stranded in the state will be given special accommodation and food. Dealer may introduce a token system to avoid crowding of card holders. Volunteers and ward members may be needed help in door delivery of grains for the elderly and bed-ridden who are unable to reach the retail outlets.
Delhi	7.2 million (including people without ration cards who need to apply online and rations to be disbursed on that basis)	Free	NSFA ration with 50% more quantity (7.5 kg rice/wheat instead of 5 kg)	1		Lunch and dinner free for night shelters.
Haryana	NFSA ration-card holders	Free	NFSA wheat & rice,* mustard oil & 1 kg sugar/hh	1		Doorstep delivery. The state Food Supplies Department has issued a list of 22 items under the PDS, adding masks, sanitizers, and gloves (no information on the list of items).
Uttar Pradesh	16.5 million construction workers, AAY card holders (poorest), destitute old-age pension holders, PWD pensioners, daily wage laborers	Free	20 kg wheat/hh & 10 kg rice/hh	2		Door-to-door delivery of essential commodities apart from the e-commerce companies and supermarkets that will use their own workforce for the home delivery of essentials; community kitchens to distribute food packets to the poor; mobile vans, e-rickshaws, thelas (carts), and other vehicles being arranged in various districts for door-to-door delivery of essential items.
Karnataka	38.3 million	Free	5 kg rice & 2 kg wheat per person	2	1 (April)	
Andhra Pradesh	14.2 million (beneficiaries already identified under the YSRCP last year)	Free	1 kg tur dal, 4 kg rice per person, oil & salt	1		Doorstep delivery by village and ward volunteers.
Telangana	8.75 million white ration-card holders, BPL, unorganized sector (beneficiaries were chosen based on a "comprehensive household survey" held last year)	Free	12 kg rice/person (double the monthly supply)	1		
Bihar	16.8 million	Free	NFSA wheat & rice*	1		In addition to food grains, Rs. 1000 DBT for food to 1.8 million ration-card-holding households. Rs. 1,000 DBT to state residents stranded outside Bihar (about 0.2 million people) because of the lockdown.
Maharashtra	NFSA ration-card holders	Free	5 kg wheat & rice	1		

Table 1 continued

STATES	TOTAL BENEFICIARIES	SUBSIDY/ FREE	QUANTITY (PER MONTH)	DURATION (MONTHS)	INSTALL-MENTS	ADDITIONAL INFORMATION
Tamil Nadu	NFSA ration-card holders	Free	15 kg rice, 1 kg dal & 1 kg cooking oil/hh	1		Through token system at an appointed day and time. Tamil Nadu follows Universal Public Distribution System.
Punjab	1 million (daily wagers and laborers)	Free	10 kg wheat, 2 kg dal & 2 kg sugar	1		Doorstep delivery. NFSA beneficiaries can collect 6 months advance ration at one go.
West Bengal	79 million	Free	5 kg rice & 5 kg wheat per person	6		
Jharkhand	NFSA ration-card holders	Free	2 months ration in advance	2	1 (April)	600 dal bhat Kendra (Rice-dal centres), 2,000 food kits (2 kg crushed rice, 1/2 kg jaggery, 1/2 kg gram).
Rajasthan	48 million NFSA ration-card holders	Free	Wheat	2		PDS food grains distribution will be carried out in phased manner to prevent crowding at these shops. Distribution of ration packs (wheat, pulses, rice, oil, and other essential commodities for urban poor not included in NFSA list).
Chhattisgarh	NFSA ration-card holders, disabled, single destitute, and Annapurna ration-card holders	Free	NFSA rice	2		
Gujarat	6 million	Free	3.5 kg wheat/person, 1.5 kg rice/person, 1 kg pulses/hh, 1 kg sugar/hh, 1 kg salt/hh	1		Distribution will be carried out in phased manner to prevent crowding at these shops.
Jammu and Kashmir	0.16 million NFSA ration-card holders	Free	NFSA rice and wheat	2	1 (April)	Doorstep delivery of PDS ration and other essential commodities.
Odisha	32.6 million NFSA ration-card holders and 0.45 million State Food Security Scheme beneficiaries	Free	3 kg rice, 2 kg wheat/person, oil	3		Three months of PDS food grains without any biometric verification. Doorstep delivery for elderly.
Uttarakhand	2.3 million	Free	NFSA wheat & rice*	3		
Mizoram	NFSA ration-card holders	Free	NFSA wheat & rice*	1		
Manipur	NFSA card holders	Free	3 kg rice, 2 kg wheat/person	1		

Source: Authors' compilation from newspaper sources through March 28, 2020.

Note: *National Food Security Act (NFSA) wheat & rice: the total quantity to be distributed per month is unclear, though the center has directed the states to allow beneficiaries to pick up 6 months' rations at one go. The center has also announced that the monthly quota of subsidized foodgrains is to be increased by 2 kg to 7 kg per person (earlier it was 5 kg per person) through ration shops for 80 crore NFSA beneficiaries. Across most of the states, there is no clarity on eligibility list of the beneficiaries. AAY/BPL/APL = Antyodaya Anna Yojana (poorest of the poor)/Below Poverty Line/Above Poverty Line. YSRCP = Youth, Labour and Farmers Congress Party.

The way forward: Getting markets to function

Past experience shows that states with more universalized coverage have lower leakages and better success at reaching the poor. The current COVID-19 context suggests that a universal transfer may be at least as effective or better than targeted transfers, as the crisis is likely to create a vast new pool of near-poor or poor households.

While social safety nets are needed, and critical to ensure food security, they cannot supersede the market, even now. Food demand comprises a variety of food products, and only cereals and some pulses are to be provided via the PDS. Many other essential commodities including edible oil, sugar, milk, salt, and pulses are needed. In addition, even if a household gets wheat, if the local millers who convert it into flour are not working, the benefit might prove useless.

The government should ensure markets keep functioning, especially by safeguarding against price gouging. The lockdown has left millions of small farmers vulnerable; they are depending on supplying fresh produce and on returns from the impending rabi harvest. Many wholesale markets (agricultural produce market committees or APMCs) initially shut down and are just beginning to open partially; many traders have been unwilling to operate, fearing infection.

It is imperative now that food move seamlessly across state borders, which can happen only if the states work together. However, the interstate transport of goods faces many problems. Truck drivers are being frisked at lockdown checkpoints, and casual laborers for loading and unloading are in short supply. In many cases, farm products are not reaching the mills. Even before the lockdown, millers had started depleting stocks of raw materials. As the wheat harvest is imminent, uncertainty looms for both farmers and traders on the status of the government's wheat procurement operations.

Disbursing aggregately, that is, providing several months' rations at once, could be the preferred mode from a social-distancing perspective. Similarly, doorstep delivery will help minimize exposure, particularly for high-risk groups such as the elderly. To the degree that the movement of food may spread the disease, moving to direct cash transfers may have public health benefits. Cash transfers may also have multiplier effects on the economy and thus provide a source of much-needed stimulus (Handa et al. 2018 or Egger et al. 2019).

The food relief program can also be used as a medium to communicate key messages about the epidemic, including on social distancing and other public health and safety measures. COVID-19 may be with us for a year or more; India's relief package is only the beginning. Going forward, the government needs a contingency plan focusing on how much, and for what length of time, the food system can continue to supply the social safety net in its current configuration. The lives and livelihoods of hundreds of millions depend on it.

This work was supported by the CGIAR Research Program on Agriculture for Nutrition and Health (A4NH), led by IFPRI.
Originally published April 6, 2020.

25. IFPRI's COVID-19 Policy Response Portal: Identifying trends and implications for food systems

Danielle Resnick

Developing countries have employed a wide range of policies to control COVID-19 and relieve economic stress. These responses continue to evolve, and different actions targeting the same problem vary widely in approach and impact. For instance, to maintain food supplies, some countries have provided direct support to farmers, some have imposed food export bans, and some do both. In addition, many responses overlap and interact, so their cumulative impacts can be difficult to interpret in isolation. For example, borders may be kept open for transporting food, but stringent screening and quarantine measures may create significant delays that deter truckers from making the trip. Providing inputs to farmers may do little for food accessibility if markets are shut down or have severely limited operating hours.

The COVID-19 Policy Response Portal (CPR) tracks these actions systematically across many different domains, enabling governments, donors, and researchers to compare policy commonalities and differences.

The CPR focuses on nine distinct types of policy responses, providing information about cross-government institutional coordination, levels of pandemic foreign aid, and citizen compliance with control measures. As some governments gradually begin to ease their lockdowns, the CPR is also tracking where policy measures have been extended or phased out. It also highlights innovations aimed at keeping food systems and livelihoods resilient.

The CPR currently includes data from 18 countries in Africa, Asia, and Europe, and its scope will be regularly expanded as new data is added.

Thus far, it reveals several important trends, including five that may significantly impact food security:

1. Major restrictions on urban food traders

2. Widespread support for contactless payments

3. Targeted support to consumer livelihoods

4. Less support for agriculture than for other forms of economic assistance

5. Exclusion of agriculture ministries in many COVID-19 national response units

Urban food traders hit hard

Urban food traders have encountered some of the most severe pandemic restrictions. Closures of open air and wet markets for extended periods, or reductions in operating hours, have been extremely common; meanwhile, many cities have banned street vendors or moved them to stationary locations. In some settings, however, authorities have disinfected markets and allowed them to continue operating, or permitted them to open every other day. In Burkina Faso, where many markets were shuttered for three weeks, food aid distributions have even been specifically targeted to informal traders. Keeping these markets open in some form is critical for maintaining traders' incomes and preventing them from sliding into poverty, and for the food security of the many urban residents who rely on them.

Support for contactless payments

Innovations in contactless forms of payment and food sourcing are proving a major advantage in coping with the pandemic, especially for traders who would otherwise need to deal with lots of cash every day. The CPR shows that many countries are reducing or eliminating fees associated with mobile money and increasing allowable transaction amounts. In Senegal, the Ministry of Trade has even launched #JaaymaMburu ("Sell me bread" in Wolof), which allows customers to order bread via an online platform and get free delivery during the month of Ramadan in order to limit queuing in front of bakeries.

Protecting consumer livelihoods

The data also show that governments are concentrating most heavily on protecting consumer livelihoods. One set of mechanisms includes fixing food prices, especially for cereals. For instance, Egypt set flour and bran prices at EGP 3,600 ($229) per ton; Mali set price limits on rice, bread, cooking oil, and sugar. Rwanda also has set fixed shop prices on staple foods. Because they focus on staples, these price controls (which bear some resemblance to those imposed during the food price crisis of 2007–2008) do less to cushion the cost of diets comprised of healthier and more diverse food options. Papua New Guinea is an exception, with mandated price controls for cereals and for eggs and fresh produce.

Agricultural support is secondary

The CPR shows that policies directly aimed at supporting agricultural production are rare compared to other forms of economic assistance. Where implemented, approaches include expanding the provision of inputs and animal feed, removing price floors placed on agricultural exports, and postponing debt payments owed by farmers. Lower visibility but higher return investments, such as agricultural research and development or extension, will likely continue to be sidelined in a time of scarce resources. In the long run, this could erode the resilience of the sector to cope with challenges such as climate change and dietary diversity.

TABLE 1 COVID-19 policy response categories and other governance categories

POLICY RESPONSES	EXAMPLES OF ACTIONS IN EACH POLICY CATEGORY (NON-EXCLUSIVE)
Restrictions on population movements	Travel bans, curfews, lockdowns at home, bans on gatherings, public transport limited, schools and religious institutions shut
Business policies	Restrictions on formal and informal markets, bans on street vending, closures of restaurants, restrictions on mining, restrictions on manufacturing, restrictions on agricultural activities, restrictions on tourism sector
Health-specific policies	Increased spending on health system, importation of test kits, drugs, ventilators, creation of new facilities for testing, increased hiring of healthcare workers
Social protection interventions	Food aid, expanded cash transfer/unemployment programs, food subsidies, expanded credit options, wage support, mandated grace period of utility bill payments, mandated grace period of rental payments
Broad fiscal policies	Nonfood price controls, support for private companies, VAT waivers for businesses, reduction of consumer taxes (e.g., sales/VAT)
Farm fiscal policies	Food price controls through procurement, food price controls through regulation, farm input subsidies, targeted rural income support
Trade policies	Export bans, export quotas, import tariffs, quantitative restrictions, quality controls
Monetary & financial policies	Exchange rate shifts, lowered interest rates, debt restructuring
Governance restrictions	Postponement of elections, state of emergency, limiting access to information, bans on political rallies, restrictions on social media platforms
OTHER CATEGORIES	**DESCRIPTION**
Price responses	Major price shifts that may occur as markets adjust to domestic and external policy changes
Citizen reactions	Protests/riots, non-compliance with population restriction measures, violent treatment of healthcare workers, destruction of health infrastructure
Institutional coordination	Government actors and agencies that are tasked with overseeing COVID-19 responses
Foreign aid	Donor commitments and actual disbursements specifically in response to COVID-19

Agriculture ministries excluded from cross-government COVID-19 response units

Finally, a unique feature of the CPR is that it shows the range of government actors involved in coordinating national responses to COVID-19. In most countries, these committees and task forces consist of health ministries working with a range of others, including commerce, industry, foreign affairs, and urban development. However, agriculture ministries are conspicuously absent. This could be due to the perception that the spread of COVID-19 is concentrated in cities and high-density areas, and that agricultural activities in rural settings are relatively safe. But this seems to be a missed opportunity to include a key sector whose reach is hardly confined to rural areas. The risk is that COVID-19 responses for other sectors may not consider possible impacts on agriculture and agrifood systems.

While policy responses to COVID-19 in some countries reflect a "copy and paste" bias, borrowing what other governments have done or resorting to common tools used in the past, other countries are charting their own course. By systematically tracking these varied responses, at both the national and subnational levels, the CPR is a resource for better understanding comparative policy processes and, in the long-term, for analyzing the combination of decisions that are most effective at protecting jobs, incomes, poverty, and food accessibility during times of crisis.

The CPR is led by IFPRI's country and regional programs and supported by Michigan State University's country programs, as well as by national institutions across the world. Funding is provided by the U.S. Agency for International Development (USAID) and the IFPRI-led CGIAR Research Program on Policies, Institutions, and Markets (PIM).

Originally published May 20, 2020.

26. Water in the COVID-19 crisis: Response, recovery, and resilience

Claudia Sadoff and Mark Smith

COVID-19 has, like nothing that has gone before, revealed the "systems wiring" of the modern, globalized world, and how destructive disturbances to those systems can be. Water is a connector across these systems, and thus has critical implications both for the effectiveness of COVID-19 response efforts and for promoting growth and building resilience in a post-pandemic world.

Water in response

COVID-19 is shining a harsh spotlight on the inequalities, hardships, and global health risks that result from the collective failure to uphold the human right to water and sanitation. In many communities around the world, a lack of water supply and sanitation deprives people of their most basic protections against the spread of the virus.

Improving water, sanitation, and hygiene has the potential to prevent at least 9.1% of the global disease burden and 6.3% of all deaths, according to the World Health Organization (WHO) report Safer Water, Better Health, released before the pandemic. Nevertheless, 4.2 billion go without safe sanitation services and 3 billion lack basic hand-washing facilities. In addition, diarrheal diseases caused by waterborne pathogens and poor hygiene inhibit nutrient absorption, so that even those with access to adequate nutrition may face malnutrition. This means that where hand-washing is limited and waterborne illness is already common, not only will COVID-19 spread more easily, its lethality could be amplified.

We should also be cognizant of the gender implications. In many parts of the world, women and girls spend hours each day fetching water or waiting in crowded queues for water vendors, potentially increasing their risk of exposure to the virus. If they struggle with these tasks because they are ill, or have to care for the sick, their health and food security could be further compromised. Compounding the issue still further, restrictions on movement may lessen the ability to access water at all.

How can we respond to these problems? In the short term, governments and international organizations should work to ensure access to safe and reliable water supplies and sanitation. This includes emergency provision for underserved communities and taking care to protect women and girls responsible for fetching water from exposure. To address potential supply disruptions, we also need a clear understanding of where and how municipal or rural water infrastructure is coping with pandemic-related spikes in demand. In Ethiopia, the International Water Management Institute has research underway now to assess the implications of mitigation measures in rural communities.

Water in recovery

Recovery from the pandemic will require effective water management that reinforces the stability of disrupted food systems. In some areas, lockdowns have impacted agricultural cycles – interrupting supplies of inputs, depressing demand, and keeping workers away from fields and factories. When farming activities resume, demand for irrigation water may rise quickly if dry season cropping expands to counter food supply deficits. Thus, a critical priority will be preparing for potentially significant unplanned irrigation withdrawals, making sure they do not undermine basic domestic water needs or overdraw aquifers, lakes, and rivers.

The risk of natural disasters – including drought, extreme weather, and flooding – occurring during the pandemic is another significant problem that threatens water security and long-term recovery. People displaced by disasters are typically relocated to densely populated camps or shelters where authorities may struggle to meet basic water, sanitation, and hygiene needs – and now, where the novel coronavirus could spread rapidly.

The prospect of overlapping shocks is yet another serious concern. The World Economic Forum 2020 Global Risks Report, published in January, ranked risks from water crises higher than either infectious diseases or food crises. In 2020, there will likely be places where we see all three at once.

To address such risks, countries will need to reinforce water governance to ensure the reliable delivery of water for priority uses, enhance water storage and irrigation capacity to head off potential crop failures and compensate for disruptions to rainfed agricultural cycles, and reduce unmanaged competition for water.

This also means better preparation for droughts or floods to mitigate the multiple shocks they can deliver to food systems. Fortunately, we can now monitor and forecast water-related risks and authorities can use those data to reduce risks of water-related setbacks to recovery, and introduce services like index-based weather insurance that will support the livelihoods of the poorest and most vulnerable people if floods or droughts strike.

Water in resilience

In the post-pandemic world, we must use what we are learning about the dynamics of these interconnected systems to "build back better." Investments in water should be used to build greater resilience to climate, health, and food system shocks, and more effective management of water-related risks.

Building back better means constructing more resilient water, sanitation, and hygiene systems that will deliver these fundamental services despite the hydrological uncertainties of climate change and growing water scarcity and pollution. It means building more "circular" water systems that secure supplies and better capture, clean, and reuse water resources in ways that protect human and ecosystem health. It means reimagining our "waste streams" as "resource streams": instead of releasing 80% of the world's wastewater back into the environment untreated, we should invest in wastewater treatment that will provide the double win of protecting communities and ecosystems against biological hazards while safely recycling water, energy, and nutrient resources. While there is currently no evidence that COVID-19 can be spread through water or wastewater, we do know that historically many epidemics have spread this way, and that untreated wastewater remains a health hazard in too many communities today.

It also means ensuring that food production and trading systems are more resilient to water challenges. To bolster domestic food supplies from COVID-19 disruptions, some countries have restricted exports and/or changed patterns of agricultural production. In addition to potential impacts on global prices, poverty, and hunger, such moves can affect water availability and undermine the resilience of food systems. Water availability and how it is allocated to multiple uses must be accounted for in food system transformations in different locations with varying geographies. Appropriate accounting for water in agricultural trade and production policies and investments is critical to sustainability. Water-scarce regions can import water-intensive crops (and their "virtual" water) from water-rich regions, where their production is sustainable and does not compete for drinking water or ecosystem requirements.

As governments and international organizations work to address these complex, overlapping challenges, systems thinking is crucial. Water connects health, food systems, climate change, nature, energy, and finance. The fabric of water security is created by weaving together effective governance, knowledge, and skills, connectivity across systems, and investment in and application of infrastructure, technologies, and services from ecosystems. The COVID-19 pandemic is stressing all of these, forcing a reckoning with many underlying problems in the process. But it is also an opportunity to expand our understanding of how these systems work and how we can build back better in a post-pandemic world.

Originally published June 4, 2020.

27. Prepare food systems for a long-haul fight against COVID-19

Maximo Torero

With a devastating one-two punch, a supply shock followed by a demand shock, the COVID-19 pandemic has knocked out the world economy. The first blow was the Great Lockdown; the second, the worst recession since the Great Depression. No modern economy has experienced anything like this. As the spread of the novel coronavirus debilitates people's ability to harvest and buy and sell food, food systems are under threat as never before.

It's thanks to producers and workers across the food supply chain that food continues to move from where it is produced to where it is needed, logistical delays notwithstanding. Countries have shown restraint, too, as most of them didn't jump to restrict food exports. In fact, the number of pandemic-related export restrictions has decreased from 18 to 7, representing less than 1% of the share of global food trade.

But the lockdowns have triggered a steep recession. The World Bank projects the global economy will shrink 5.2% this year. The IMF's latest projection is -4.9%. The OECD is forecasting a 7.6% contraction, given a second wave of infections before the year's end, and a very slow recovery of 2.8% in 2021. In both rich and poor countries public debt is soaring and is expected to exceed the post-World War II peak.

A collapse in demand for food due to lack of income and disruptions to local food markets indicates important vulnerabilities, as shown in Figure 1, and could prompt a global food crisis. Hunger and malnutrition were significant global problems even before the pandemic. More than 2 billion people didn't have regular access to safe, nutritious, and sufficient food last year. Some 704 million of them went to sleep on empty stomachs; this included 135 million people who were on the edge of starvation.

At FAO, we estimate that a 5% to 10% drop in GDP growth would mean an additional 38.2 to 80.3 million people in poor countries that rely on food imports falling into the hunger trap. At a global scale, this means the number of hungry people would jump by between 74 million and 120 million. The effects of COVID-19 are even more pronounced in Africa south of the Sahara and small island developing states (SIDS). The virus has shuttered tourism, leaving SIDS such as Fiji, the Maldives, and Mauritius scrambling for economic survival. The World Bank expects to see a sharp 20% drop in global remittances.

Africa is bracing for the worst. The epic oil price crash has led to a global financial bust. For the continent's exporters, such as Nigeria, Chad, Libya, and Algeria, it has wiped out their principal source of revenue. A catastrophic locust outbreak in East Africa was – pre-pandemic – projected to force 25 million people in Ethiopia, Kenya, Somalia, Sudan, and Uganda to go hungry. A swarm covering

FIGURE 1 Structural vulnerability and known food insecurity hotspots

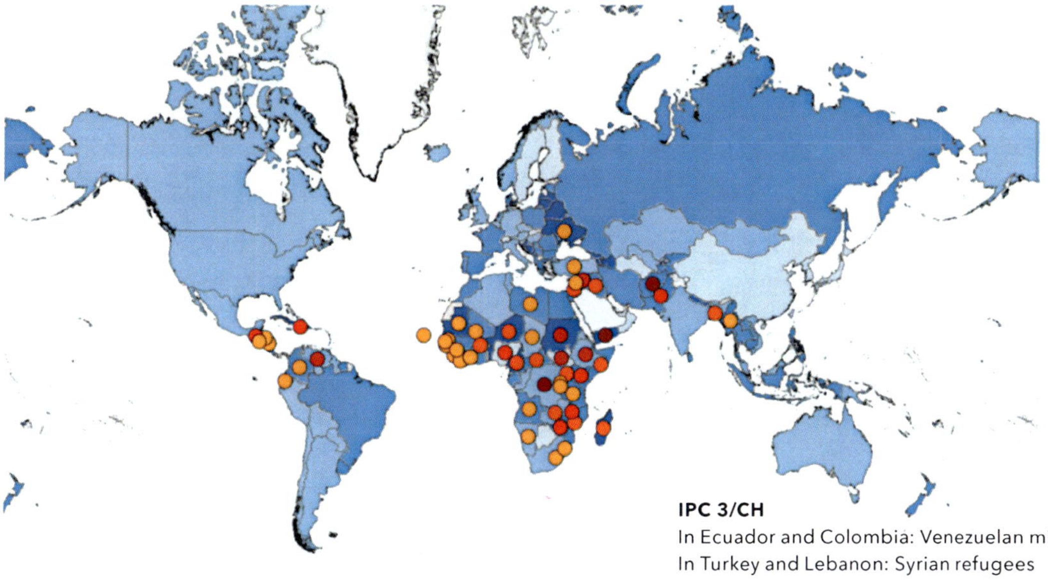

Source: FAO/Hand-in-Hand, Integrated Phase Classification.

Note: IPC/CH (Integrated Phase Classification/Cadre Harmonisé) provides a scale for classifying the severity and magnitude of food insecurity and malnutrition. Phase 3 (IPC 3) indicates crisis-level food insecurity.

FIGURE 2 Formal jobs at risk in food systems

Where in value chain	Jobs (in millions)	Livelihoods (in millions)
Primary production	716.77	2,023.80
Food processing	200.73	484.54
Food services	168.97	339.44
Distribution services	96.34	241.48
Transportation services	41.61	101.05
Machinery	6.51	13.18
Inputs	4.89	11.06
R&D	0.13	0.29
Total	1,280.93	3,214.84
Total at risk due to COVID-19	451.64	1,090.89

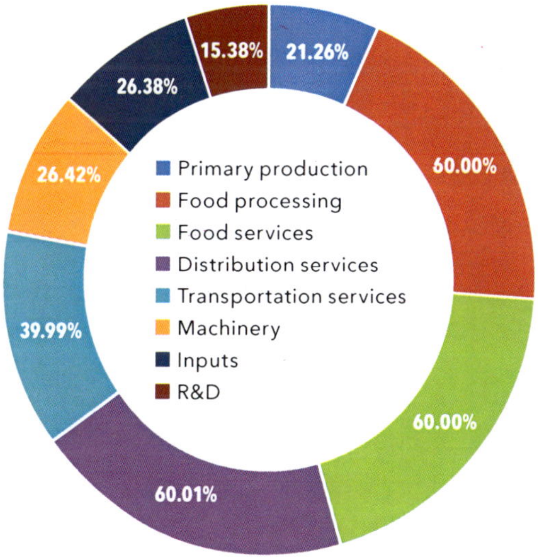

Source: FAO/IFPRI unpublished estimates, based on ILO 2020 – ILO extrapolation scenario. Not annualized. Jobs represent formal employment; livelihoods cover a broad array of self-employed, informal, migrant, and seasonal labor.

one square kilometer contains 80 million insects that consume more food in 24 hours than 35,000 people. Africa south of the Sahara faces its first recession in 25 years and is especially vulnerable to the impact of COVID-19.

Food systems, which directly employ over a billion people, are about to lose more than 451 million jobs or 35% of formal employment, according to an unpublished FAO/IFPRI estimate. The jobs most at risk are in food processing, services, and distribution, disproportionately affecting female workers.

Countries must respond by deploying the full power of fiscal and monetary policies. I cannot overemphasize the importance of expanding social protection for vulnerable people who can't afford basic nutrition. Governments should use cash transfers and mobilize food banks. Parallel to this, they should increase food production, reduce food losses, and create employment. Public works projects throughout agrifood systems can provide people with livelihoods. It is important that the rural poor, especially the women among them, benefit from this policy combination.

Food supply chains must keep moving. This means protecting the health of all supply chain workers. Economic recovery cannot come at the expense of health, as seen in meat processing plants in the United States and Germany, and wholesale markets in Mexico, Peru, and Brazil. Health is a precondition for economic recovery; and food is a precondition for health. Similarly, there is a need to increase testing capacity at ports to allow vessel crews to disembark without the need to self-quarantine and minimize disruption to maritime transportation.

It is equally critical that smallholder farmers and micro, small, and medium-sized enterprises (MSMEs) keep operating. In poorer countries, they play a crucial role in supplying food to poor consumers. Supply chain disruptions have hit MSMEs hard, and they need access to finance to stay afloat during the drawn-out period between recession and an upturn in a U-shaped recovery. Central banks or international financial institutions should provide warranties, so that banks can help MSMEs with highly concessional emergency loans, business continuity grants, and moratoriums on loan repayments, as well as short-term stimulus packages that support sales, cash flow, and working capital. Banks should set lending targets for smallholder producers and engage in inclusive agricultural investment.

If small enterprises in agricultural value chains shut down, the problems of food access and food availability could intersect, creating a nightmare scenario the world is ill-equipped to handle.

Finally, countries have to accelerate intraregional trade. Exports can mitigate losses in revenues. And imports can improve food availability and stabilize local food prices. In both exporting and importing countries, access to various markets can boost producers' productivity and income.

For Africa, trade within the continent is especially important, because the region can create demand to compensate the weak demand from Europe. African countries should develop food safety standards across the value chain and ramp up access to infrastructure. The first is vital, as it would reduce nontariff trade barriers and prevent governments from imposing blanket import restrictions.

COVID-19 has amplified the voices of antiglobalization. It is setting off calls for food self-sufficiency as well. It's understandable, but pursuing food self-sufficiency is the worst move countries can make right now. No country has all the natural resources to produce the food it needs in the variety it needs. Facilitating global trade, not promoting self-sufficiency, is key to boosting food security. The pandemic has also given us an opportunity to make investments that will lay the foundation to reset food systems and whose returns will accrue far into the future.

The world is in a grueling 12-round fight against the coronavirus. In every round, there's a risk of another lockdown. But the above policy recommendations will prevent millions of people from facing outright starvation. They will make it easier for countries to bounce back from the recession and go the distance.

Originally published July 2, 2020.

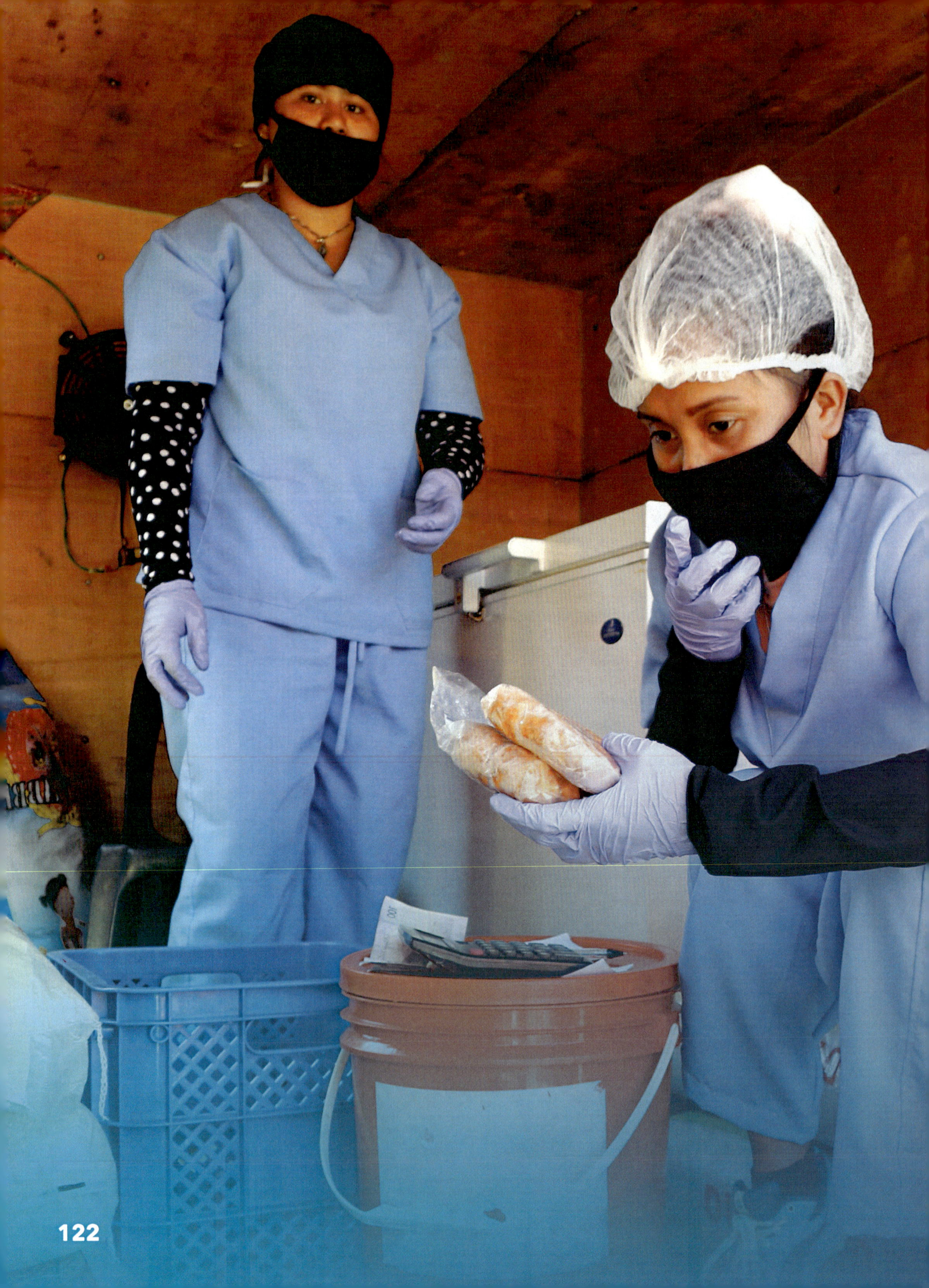

THE FUTURE OF PANDEMICS AND FOOD SYSTEMS

28. Africa's growing risk of diseases that spread from animals to people

Bernard Bett, Delia Randolph, and John McDermott

Three-quarters of emerging human infectious disease outbreaks are "zoonotic," meaning they originate from viruses and other pathogens infecting animals that then "jump" species to infect people. This "species jump" by pathogens is not new – it has occurred throughout pre- and recorded history. But in the last half of the last century, with the widespread use of antibiotics and vaccines, many had begun to believe that the era of infectious disease was ending. The story of epidemics, however, is always evolving. As we see clearly now with the ongoing COVID-19 pandemic, which is believed to have originated from virus-infected meat or live animals sold in a traditional "wet" food market in Wuhan, China, our hopes for the end of infectious disease were badly misplaced. Over the last 100 years, in fact, there has been growing evidence of not less but more frequent emergence and greater spread of zoonotic pathogens in humans and animals. In recent decades, most of these zoonotic pathogens were reported in Europe and the United States. More recently still, Asia, Africa, and South America appear to be growing in importance as origins of zoonotic pathogens.

For centuries, East and Southeast Asia have been the hotspots of influenza and other emerging zoonotic diseases with pandemic potential, but in this century the region has also been the origin of novel coronaviruses causing both the 2002–2003 epidemic of severe acute respiratory syndrome (SARS) and the 2019 coronavirus disease dubbed COVID-19. A major cause of the emergence of new influenzas is the increasing densities of people and their domestic animals. Greater human populations are also increasing human interactions with wild animals, which is speeding the acquisition of disease infections among people.

Africa is now catching up to Asia as an infectious disease hotspot. Africa now has the fastest-growing and youngest human population of any region in the world. In 1900, Africa south of the Sahara had around 100 million inhabitants; the population now stands at 1 billion and by 2100 is projected to grow to around 4 billion people. With increasing human populations and increasing demand for milk, meat, and eggs due to rising urbanization and incomes, the densities of humans and domestic animals are also increasing – particularly in coastal West Africa and North Africa and the highlands of East Africa. Figure 1 compares the current human, poultry, pig, and ruminant populations across Africa and Asia. Some regions of Africa are now approaching the high density levels seen in Asia.

In past centuries in Africa, animal pathogens jumping to humans almost always caused limited outbreaks – reflecting the comparatively low densities of people and animals and their relative isolation. However, this pattern is changing, with increases in both frequency of emergence and expanded spread in human populations. Here, we highlight key changes in human, animal, and environmental health drivers contributing to more frequent emergence and greater spread of emerging zoonoses in Africa, now and in the future. Understanding these changes is critical in developing preventive and rapid response strategies and capacities to mitigate the increasing risk of epidemics of emerging diseases in Africa.

FIGURE 1 Population densities of humans, poultry, pigs, and ruminants in Africa and Asia

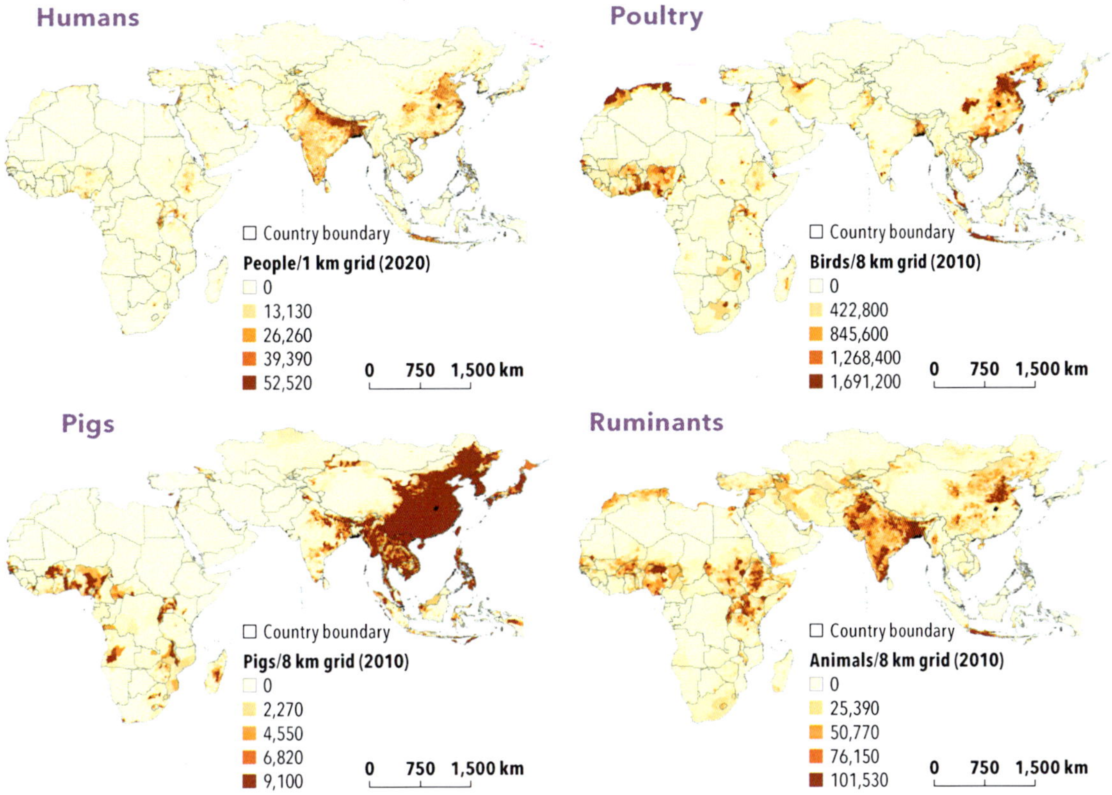

Source: Human population data were obtained from World Pop; livestock data were obtained from the Gridded Livestock of the World database, composed by Fred Otieno, ILRI.

Note: Regions that have high human populations in Africa include East and West Africa; in Asia, relatively high human populations occur in southeast China and India. In general, areas with high human populations also have high poultry and other livestock populations.

Emergence and spread of zoonotic pathogens follow different patterns

While there are commonalities, each outbreak, epidemic, and pandemic has its own unique features. Tracing pathogen emergence from one host species to another has been greatly aided by the advent of genomic tools and improved but still limited sampling of the host species. These methods have helped us better understand the movement of pathogens from primates (HIV-AIDS), bats (Ebola), and rats (Lassa fever) to humans. Zoonotic pathogens can directly jump from an animal species to infect humans (HIV-AIDS from primates) or through other animal species that either act as an intermediate connector host or bridge (SARS-coronavirus and SARS-coronavirus 2 that causes COVID-19, from bats through other wildlife species then to humans) or as amplifier hosts of pathogens transmitted to humans (Nipah virus from bats, multiplied in pigs; influenza viruses mixing between human, pig, and poultry populations in East and Southeast Asia). While many new diseases originate in wildlife, for some of the most serious, livestock have been a connector or amplifier host.

Prevention or, failing that, rapid initial containment before an exponential growth of cases is the health goal. Low population density and stable societies serve as natural preventive measures. In Africa in past centuries, infectious pathogens jumping from animals to humans almost always caused limited outbreaks or "burned out." For example, simian immunodeficiency viruses have likely been transmitted from primates to humans from prehistoric times, but did not cause serious epidemics until the late 20th century. But the dramatic social, demographic, and health changes that began in late 19th century Africa helped to transform these occasional pathogenic wildlife-human spillovers into pandemics of human-to-human disease transmission, such as the human immunodeficiency virus (HIV).

This new pattern of disease emergence is unfortunately likely to become increasingly common, given the dramatic rise in Africa's human population.

TABLE 1 Important examples of recent epidemic zoonoses not previously known

EMERGING ZOONOSIS	PRIMARY ANIMAL HOST	AMPLIFYING ANIMAL HOST	GEOGRAPHICAL IMPACT	APPROXIMATE DATES
HIV-AIDS	Primates		Global with major burden in Africa	Late 1970s to present
Ebola	Bats*	?	Africa (Central, East, West)	Varied outbreaks; major epidemic in 2015
Nipah	Bats	Pigs	SE Asia	
Severe acute respiratory syndrome (SARS)	Bats	Civets	Origin in China to multiple other countries	2003
Middle East respiratory syndrome (MERS)	Bats	Camels	Middle East and East Africa	2008 to present
Covid-19	Bats	tbd	Origin in China to global	December 2019 to present
Avian flu (H5N1)	Wild birds	Poultry	East/Southeast Asia to global (Americas relatively spared)	2005 to 2010
Swine flu (H1N1)	Pigs		Global	2009

Source: Authors' notes.

Note: *Bat transmission of Ebola is assumed but not confirmed.

What might change regarding the frequency of pathogen emergence in primary animal hosts and subsequent transmission to humans and domestic animals?

This zoonotic pathogen and human disease pattern continues to evolve and change. Infectious zoonoses producing severe clinical illness and high mortality, such as Ebola and HIV-AIDS, are the most highly visible signs of emerging zoonoses in Africa. As in Asia, increasing changes in land use, including the expansion of human settlements and agricultural lands, are increasing contacts between humans and wild animal host species. That human disease outbreaks of yellow fever and other hemorrhagic fevers are associated with exposure to new pathogens through human incursions into forests has been well known for two centuries. But the routine exploitation of forests for mining and other resource extraction purposes in recent decades has created new opportunities for viral transmission. Rising human populations in parts of Africa are accelerating the use of forests for hunting bushmeat for consumption and use in traditional medicines and trade. As in China, wet markets for bushmeat are also found in Africa and there is also considerable illegal international trade.

Closer interfaces between dense human settlements and forests continue to expand. Many wildlife species – most worryingly, from a pathogen emergence perspective, bats – are increasingly adapting to peri-urban living. Accra, the capital of Ghana, is home to more than a million fruit bats and hunting and sales are important economic activities. One critical question is whether bat pathogens, including a range of bat coronaviruses, are also evolving to become more adapted to multiple animal hosts, including humans. Analyses of bat coronaviruses, including SARS-Cov-2, indicate that they may more easily mutate to infect humans than in the past.

What might change in the frequency of zoonotic pathogen emergence and spread from intermediate or amplifier animal species?

For some emerging zoonoses, the spread of infections in domestic animal species is the key factor. For both Nipah and MERS (Middle East respiratory syndrome, caused by a coronavirus), the spread from bats to intermediate domestic animal hosts was important in the subsequent emergence of the disease in humans. These zoonoses are among those that continue to exist and could "re-emerge" at higher rates in humans with increasing human densities and poor management and hygiene of their main amplifier species – pigs for Nipah and camels for MERS.

Influenza viruses have been responsible for many epidemics over the past centuries, including the flu pandemic of 1918-19 that killed more people than any other documented pandemic – one-third of the world was infected and around 50 million people died. Interestingly, in the H5N1 (avian flu) epidemic of 2006-2011 that raised global animal and human health concerns, H5N1 infections became endemic in Indonesia but were relatively quickly eliminated following their introduction in West Africa, perhaps because Indonesia at that time had much greater poultry densities than West Africa. But West African poultry density is now catching up to Asia's.

What is changing in human societies that facilitates the spread of emerging zoonotic pathogens?

Increasing human populations, urbanization, and rising incomes are changing Africa in fundamental ways. One major change is a dramatic increase in air travel between Africa and the rest of the world. While traditionally most African travel connections have been through Europe or the Middle East, in the past decade the number of African connections to Asia has been rising. The first "African" case of COVID-19 was diagnosed in Nigeria in a traveler coming from Italy.

As in other developing regions, Africa's health and social support systems to serve the growing and more mobile human populations have lagged. This emerged as a critical issue in the AIDS epidemic in Africa. With regard to COVID-19, Marius Gilbert and colleagues combined data on Chinese air travel connections with available indices on health preparedness and infectious disease vulnerability indices to rank country risk. As noted above, Ebola outbreaks in Africa have usually been contained locally; the 2015 West Africa Ebola epidemic spread through countries with very weak health systems. In 2019, an Ebola outbreak in the Democratic Republic of the Congo persisted as it occurred in a conflict zone. The combination of Africa's weak health systems, the expanding health needs of its growing populations, and its ongoing conflicts are a great concern for the continent's emerging zoonoses preparedness and response.

Implications and opportunities for controlling emerging infectious diseases in Africa

The rising risk of emergence and spread of zoonoses in Africa has significant consequences for the continent and the rest of the world. Epidemics in recent decades have varied in both their causes and effects and there are no common guidelines for the prevention or early control of zoonotic diseases.

To increase Africa's resilience to the threat of emerging zoonoses, regional and global cooperation are essential. The continent's disease control capacity and preparedness programs should be increased and scarce resources should be transferred to where they are needed most. These require strengthening regional human (WHO regional office for Africa) and animal health (African Union-InterAfrican Bureau for Animal Resources) bodies. Governments and organizations should also adopt a coordinated One Health response across human, animal, and environmental health. Bringing these three disciplines together is essential to respond to the increasing threat of emerging zoonoses in Africa.

The record thus far on COVID-19 and on past disease outbreaks shows that early, effective, and sustained response is essential to winning the battle over these diseases. Innovative use of information and communication tools and platforms and engagement of local communities are crucial to improved disease surveillance and effective response. Building these systems requires demand from the public and commitment from policymakers and investors. COVID-19 is a game-changer. It has shocked the world and continues to disrupt the daily lives of billions of people. Its eventual impacts on Africa are not yet fully apparent but may be enormous. But it will also provide important lessons in disease prevention and early response, the kind of lessons routinely ignored in the past, and point the way to combining effective disease-fighting practices linking human, animal, and environmental health.

This work was supported by the CGIAR Research Program on Agriculture for Nutrition and Health (A4NH), led by IFPRI.
Originally published April 7, 2020.

29. COVID-19 and the promise of food system innovation

Corinna Hawkes

One of the silver linings of any crisis is the innovation it produces. And when it comes to food, COVID-19 is no exception.

The evidence, anecdotal as it is, shows that lockdowns around the world have had a profound impact on the markets, transport, and labor supply needed to produce, distribute, and sell nutritious foods. With reports of vegetables rotting in the fields and milk being thrown away while people go without, a clear mismatch has emerged between supply and demand.

But as things have closed down, new spaces have opened up. Innovations driven by government, business, and communities targeting production, distribution, markets, and consumers have proliferated to enable food to get to people who need it in new ways. Crisis often necessitates new actions for short-term solutions. But given the longer-term problem of undernutrition and overweight around the world, and the ongoing conversation about how food systems need to change, it's worth asking: do these innovations tell us anything about what is possible and beneficial for food systems transformation toward nutritious, healthy diets for all?

Let's look first at the innovations themselves. Digital innovations have been noteworthy in enabling producers to conduct their businesses in new ways. In China, where the pandemic began, the Chinese Agricultural Product Market Association has been working with e-commerce businesses and mobile chat groups to provide online platforms to help supply meet demand; suppliers post information about what foods they have and buyers do likewise about what they need. Government has also simplified registration procedures and provided training on sales and consumer-oriented marketing to make it easier for farmers to set up e-enterprises for their products. In India, the National Informatics Centre created the Kisan Rath mobile app to help farmers and traders find vehicles to move their fruits and vegetables to market. In Malawi, farmers are reportedly adding value to products otherwise lost – tomatoes into pastes, for example – and using online advertising platforms to get the word out. In Oman, the Ministry of Agriculture and Fisheries has established an online auctioning platform to enable electronic bidding for fish.

Innovations in distribution have likewise been driven by government, as well as by communities themselves. In India, an amendment to the Agriculture Produce Market Committee Acts now allows farmers to sell their harvests from multiple locations and to any buyer, rather than just in designated markets. In Fiji, the Agriculture Marketing Authority stepped in to buy fresh foods direct from suppliers unable to travel to market, selling them on at no added cost to market vendors. In Nepal, communities established "agri-ambulances" to get vegetables from farm to market.

Elsewhere, there have been innovations in point-of-sale. In Quito, Ecuador, sales from the city's extensive network of urban and peri-urban agriculture have been diversified to include third-party transactions and basket sales direct from the gardens. In Addis Ababa, Ethiopia, the government has provided a renovated stationary bus as a venue for urban farmers to provide direct market access to local people.

For people going short on food, new methods are likewise being used to supply food, vouchers, and meals. Innovations at the urban level are of particular note. From the food deliveries in Lima, Peru, to the community kitchens in Freetown, Sierra Leone, and Masiphumelele, South Africa, communities, governments, private enterprise, and funders are finding new ways to feed people. Other examples include vouchers for families formerly supported by the nationwide free school meals program in São Paulo, Brazil, and repurposed buses being used to deliver food in Wuhan, China. At the national level, it is reported that governments in the Pacific Islands and Sri Lanka are distributing seeds to encourage households to grow their own. Much more dominant are new social protection measures: the World Bank reports that as of June 12, 2020, 173 countries had enacted 621 new social protection measures, including cash transfers and in-kind food and voucher schemes – a vital lifeline to enable people in poverty to afford nutritious foods. Some have also taken measures to stabilize prices. Sri Lanka, for instance, is reported to have fixed the wholesale price of vegetables. New finance measures have been taken with potential to support small food enterprises and food production.

Needless to say, it's not known if these innovations are working. Doubtless, some are not. Others will be mired in political conflict; plenty more will not go far enough. Others may even be inappropriate – such as reports of junk food in delivery schemes – or misplaced. Elsewhere, doors to novel ideas remain firmly shut. So this is not a case of championing innovations for innovations' sake, and endorsing them without a critical eye. Rather, it is a case of assessing what can be learned from them about what is possible and beneficial for food systems change.

Here are three ways these innovations show the way forward. There are likely more.

First, food systems solutions to ensure the right kind of food gets to those most vulnerable are possible. During COVID-19, the bureaucratic, financial, logistical, and technological reasons that always seemed to make actions impossible or improbable have fallen away. This shows, at its heart, that it is a political choice of whether to act or not. When there is a will, change is possible.

Second, concerted, creative, and cross-sectoral intervention is needed to get food systems working for better diets. It's not something that can simply be left to happen without a clear plan. Important as government is in these interventions, innovation also needs to involve communities, businesses, and partnerships. Creative thinking is needed to find the right solution from the diversity of possible innovations; this is not the time to fall back into pre-held assumptions about how food systems ought to work. Solutions can come in the form of hard regulation, for example, as well as business-driven solutions, through local markets as well as global ones.

Third, innovation is a huge opportunity to build evidence for the way forward. COVID-19 has provided a real-life innovation lab, a testing ground for big ideas. Test, fail, succeed, learn, change. What is working (or not) and why? What can we learn from this to redesign food systems? Experimenting our way to the future can and should be a way forward. A next step should be to assess what can be learned about what works (and what does not) and which innovations show most promise in effecting food systems change at different levels.

COVID-19 has disrupted food systems everywhere. But it has also provided an unprecedented opportunity for innovation, a space in time when immediate needs have spurred responses never seen before, a base on which to redesign food systems for the better.

Originally published June 18, 2020.

30. COVID-19 and resilience innovations in food supply chains

Thomas Reardon and Johan Swinnen

The COVID-19 pandemic has triggered intense discussions about the vulnerability of the world's food systems and food supply chains (FSCs) and about the roles of different types of supply chains, such as local vs. global, in providing food security. We know that the spread of the novel coronavirus and government-imposed lockdowns and other restrictions have had a range of impacts on FSCs, and triggered a variety of creative innovations to keep supply chains running.

To guide government policy responses going forward, and to facilitate a shift to more resilient FSCs in the long run, we need to understand several things: the role of various types of supply chains in food security; how resilient they have – or have not – been to the pandemic's impacts; and what innovations are now emerging to improve their resilience.

Here, we distinguish between global chains (where the food or agricultural raw material is produced in one country and consumed in another) and domestic chains (where food is produced and consumed in the same country). Within domestic chains, it is useful to distinguish between those relying on small and medium-sized enterprises (SMEs) in logistics, trade, processing, and retailing; and those dominated by large-scale enterprises, including fast food chains, supermarkets, large processors, and big logistics firms.

While there are obviously important differences across commodities and countries, available data suggest that domestic supply chains, especially those dominated by SMEs, are by far the most important for supplying food to consumers in developing countries. Rough estimates suggest that, on average for South Asia and Africa south of the Sahara, domestic chains account for between 75% and 90% of food consumed, of which the vast majority comes through SME-dominated chains and up to 20% through large-scale enterprises. Global chains account roughly for 15% to 20% of food consumption in these regions, with a positive correlation between GDP and their share.[1]

Pandemic-related disruptions in supply chains are concentrated in their labor-intensive segments. In general, supply chains in rich countries have been more resilient because they are more capital- and knowledge-intensive. Notable exceptions are harvesting that depends on migrant labor; labor-dense processing such as in meat processing in the United States; and obviously restaurants and other food-service sector firms.

Still, there are important differences among FSCs in developing countries. Global FSCs have been more resilient because trade is mostly undertaken by large enterprises in coordinated and capital-intensive supply chains that can mostly adjust to disruptions geographically and temporally, and somewhat in product composition. While there is much concern about COVID-19 affecting trade in

1 Barrett, C., T. Reardon, J.Swinnen, and D. Zilberman, 2020, "Agrifood Value Chains Revolutions in Low and Middle Income Countries," unpublished manuscript. See appendix.

perishables, most extraregional trade is organized through large capital-intensive firms.[2] These large trading companies can reduce risk and adjust to shocks as they are more flexible in switching global sourcing and destination regions and in diversifying and shifting stocks to manage risk – as they already do to manage risks from climate shocks (Reardon and Zilberman 2018).

Within domestic FSCs, COVID-19 and lockdowns have mixed effects. Large-scale companies are generally less labor intensive but rely more on hired labor (affected especially by lockdowns), while SMEs are more labor intensive, but use more family labor. Wholesaling and logistics operations, such as third-party logistics firms (3PLS) in trucking and transport, which are very important for food transport in Africa south of the Sahara, are disrupted by mobility restrictions and wholesale market restrictions. These also affect farm input distribution in rural areas. These differences matter for processing, trade, and logistics, and also apply to the farm sector. Larger mechanized farms are less affected by pandemic restrictions, but those that depend on hired labor have felt an impact. Hired farm labor is relatively rare in Africa south of the Sahara, except for labor-intensive poultry and horticulture operations, compared to India, for example, where farms depend much more on hired labor (Reardon et al. 2020).

Supermarkets and large processors in developing countries depend largely on SME wholesalers, but the largest companies – such as Future Group, a leading supermarket chain in India – tend to have their own logistics and procurement units. This allows them more control and coordination to maximize their sourcing in the face of constraints. SMEs have to take what they can get.

Innovations and food chain resilience

Supply chain participants have introduced a series of innovations in response to COVID-19 and restrictions. So far, there is little systematic data available on these entrepreneurial and creative institutional responses. Anecdotal information suggests that these innovations are important and could have a major impact on the future of FSCs in developing countries – in particular buttressing their resilience. Entrepreneurs are telling us that "what we thought would only be possible over the next two decades is now being introduced in a few months."

There are several kinds of mutually supporting innovations:

Social innovations. Labor interactions in value chains are being reorganized to reduce shortfalls of access to labor, while guaranteeing worker safety. This involves:

- Increased flexibility of labor sourcing and timing, including facilitating the movement and safety of workers. For example, in Nigeria large chicken processors are busing workers to plants and increasing the number of shifts so there are fewer workers in the plant at one time.

- Increased flexibility by replacing workers with machines. This is easier for large farms and large firms, as they are more mechanized to begin with. This of course implies a challenge of reduced employment both in the short and long term.

2 For example, Bakhresa, the biggest food processor in East Africa, imports wheat that then goes to big mills; or see activities of companies like Charoen Pokhphand Food in Asia (see AGRA 2019 and Reardon and Zilberman 2018).

Business strategy innovations in systems of input procurement and output or service marketing. The business adjusts its systems to mitigate risks such as a rapid drop in demand among its usual clientele, or a sudden blockage to its sourcing a key input. The changes are parallel to those made for labor:

- Increased flexibility in marketing by diversifying the customer base and ways to reach consumers. For example, restaurants quickly moved from on-site service to delivery (as we discuss more below). Likewise, a supplier might have previously targeted only food service and then shifted to retail or direct sales to consumers.

- Increased flexibility in sourcing by diversifying logistics; diversifying input types to get what's available; and diversifying geographic sources to reduce risk. The latter resembles actions firms were taking pre-pandemic to address climate risk (Reardon and Zilberman 2018).

Technological innovations. Introducing technologies that improve hygiene while requiring fewer personal interactions between workers, and between the firm and customers. Examples include contactless delivery and e-commerce for customers.

Financial resilience innovations. Large companies are also creating financial resilience innovations for SMEs. In India, Swiggy, a fast-growing food delivery app and logistics company, delivers for 40,000 restaurant partners, helping them with its "jumpstart package" to recover sales, while the Swiggy Capital Assist Programme helps pay for hygiene and distancing upgrades. In Singapore, Unilever Foods Solutions partnered with e-commerce platform Carousell to launch #SupportLocal, enabling 180,000 food and beverage firms in Southest Asia to connect online with local diners. Unilever also shifted to advance payments to small farmers and credit to small retailers to support their resilience over the past three months.

Growth of e-commerce

E-commerce is a particularly vibrant example of innovation. While the use of e-commerce in most developing countries has generally been low, in some it was growing rapidly even before COVID-19. For example, in China – with more than a billion people now online – e-commerce was increasingly widespread, even in rural areas.

Apart from COVID-19, the demand-side drivers of e-commerce are similar to those of the "supermarket revolution": increasing opportunity costs of time for shopping, magnified by traffic time with urban congestion, enhances the benefits of one-stop-shopping at supermarkets – and now e-commerce.

The supply-side drivers are (1) rapid diffusion of digital technologies, internet, computers, and mobile phones; (2) intense competition and investments in the past decade by e-commerce multinationals (first Amazon, then also Alibaba) joined by e-commerce domestic firms (such as Flipkart in India and Jumia in Nigeria); (3) e-commerce by supermarket chains (such as the Walmart-Flipkart joint venture in India); and (4) complementary investments by logistics firms (such as FedEx and local counterparts), delivery firms (such as Instacart and Deliveroo), and mobile money firms.

COVID-19 has accelerated the first wave of e-commerce diffusion already underway led by large companies, and – encouraged by governments and NGOs facilitating e-commerce platforms – created a second wave into the realm of SMEs in trade, logistics and delivery, and mobile money firms.

E-commerce is growing fastest in Asia, but is increasingly spreading in Africa too. Large e-commerce companies are rapidly developing both retail services and intermediation services to help SMEs. In response to COVID-19, Alibaba has scaled up local deliveries of fresh produce to Chinese consumers. It adapted its online shopping site Taobao to provide deliveries in "one hour" with "hyperlocal fulfillment" including from SME retailers and independent chains (Song 2019; Chou and So 2020). In India, Flipkart is growing fast during the COVID-19 crisis and developed a "hyperlocal delivery" grocery service linking SME suppliers with domestic supermarket chains like Vishal Mega Mart with its e-commerce operations (Economic Times 2020, 2020). In Nigeria, Jumia has seen its year-on-year sales quadruple with COVID-19 (Kazeem 2020).

SMEs are also starting their own e-commerce services to cope with COVID-19. In Thailand, SMEs are selling food directly to consumers via Facebook and local delivery apps over mobile networks (Leesa-Nguansuk 2020). Malaysia-based MyFishman.com provides fresh seafood subscriptions and delivery services to local fishermen (Harper 2020).

Business associations and governments are also facilitating e-commerce during COVID-19. In China, the China Agricultural Wholesale Market Association began working with e-commerce and mobile chat groups to link suppliers and buyers (Fei and Ni 2020). In Myanmar, the Myanmar Pulses, Beans & Sesame Seeds Merchants Association started an e-platform to link domestic suppliers and processors and exporters.[3] In India, the National Informatics Centre created the Kisan Rath mobile app to help farmers and traders find vehicles to move their fruits and vegetables to market (Financial Express Online 2020).

Lessons and recommendations

The resilience of domestic supply chains is crucial to food security in developing countries. They are dominated by SMEs that have been particularly vulnerable to pandemic-related impacts, and to a lesser degree by emerging large enterprises and global value chains, both somewhat better equipped to weather COVID-19 shocks. Domestic FSCs have been particularly disrupted downstream in food service and retail, moderately in processing, and much less so in farming, except where hired labor is important.

Steps by the private sector – both large firms and SMEs – include introduction of flexibility in labor access, in product procurement, in marketing, in technology, and in financial resilience. In many cases the innovations by large firms, such as with e-platforms and credit, have made SMEs' suppliers and retailers more resilient. The expansion of e-commerce has accelerated, and we expect that to continue post-pandemic. E-commerce has helped SMEs deliver food to consumers under lockdowns and other constraints, and added to the resilience of the supply chains in developing regions.

3 We are grateful to Curtis Slover of UNOPS for this field observation.

Many government officials and donors have worried that the pandemic would simply stop the operation of supply chains – requiring them to step in and replace the market. This would be neither possible, given the massive scale of the market and food demand, nor necessary in many cases, given the steps supply chain actors are taking to adapt and build resilience. Of course, these efforts have not been universally successful – there is obvious evidence of real dips in business activity and demand, with accompanying employment losses. But the many cases of innovation we identify paint a picture of a private sector, as well as associations and governments, keen and able to innovate. Governments and development partners would do well to support that innovation with investments in hard and soft infrastructure and an enabling business and commerce environment for both SMEs and large companies eager to play their part in resilience for food security during the pandemic and recovery – and in rebuilding for the future.

Thomas Reardon thanks the United States Agency for International Development (USAID) under: (1) the Feed the Future Innovation Lab for Food Security Policy Research, Capacity, and Influence (PRCI), and (2) the Feed the Future Sustainable Intensification Innovation Lab (SIIL).

Originally published July 6, 2020.

Contributors

Johan Swinnen is director general of IFPRI, Washington, DC, USA.

John McDermott is the director of the CGIAR Research Program on Agriculture for Nutrition and Health (A4NH), Washington, DC, USA.

Channing Arndt is director of IFPRI's Environment and Production Technology Division, Washington, DC, USA.

Marc F. Bellemare is a professor in the Department of Applied Economics at the University of Minnesota, St. Paul, MN, USA.

Bernard Bett is a senior scientist, Animal and Human Health, at the International Livestock Research Institute (ILRI), Nairobi, Kenya.

Ruchira Boss is a research analyst for the CGIAR Research Program on Agriculture for Nutrition and Health (A4NH), New Delhi, India.

Antoine Bouët is a senior research fellow in IFPRI's Markets, Trade, and Institutions Division, Washington, DC, USA.

Clemens Breisinger is a senior research fellow in IFPRI's Development Strategy and Governance Division and head of IFPRI's Egypt Strategy Support Program (ESSP), Cairo, Egypt.

S. Mahendra Dev is vice chancellor of the Indira Gandhi Institute of Development Research, Mumbai, India.

Alan de Brauw is a senior research fellow in IFPRI's Markets, Trade, and Institutions Division, Washington, DC, USA.

Kevin Chen is a senior research fellow with IFPRI's East and Central Asia Office in Beijing and chair professor at Zhejiang University, Beijing, China.

Eugenio Díaz-Bonilla is head of IFPRI's Latin-American and Caribbean Program, Washington, DC, USA.

Eve Dill is a social science research professional with the Rural Education Action Program (REAP) in the Freeman Spogli Institute for International Studies at Stanford University, Stanford, CA, USA.

Xinshen Diao is the deputy director and a senior research fellow in IFPRI's Development Strategy and Governance Division, Washington, DC, USA.

Shenggen Fan is a chair professor at China Agricultural University, Beijing, China, and former director general of IFPRI.

Sherwin Gabriel is a scientist with IFPRI's Environment and Production Technology Division, Pretoria, South Africa.

Daniel Gilligan is deputy director of IFPRI's Poverty, Health, and Nutrition Division, Washington, DC, USA.

Joseph Glauber is a senior research fellow in IFPRI's Markets, Trade, and Institutions Division, Washington, DC, USA.

Yetimwork Habte is a research officer with IFPRI's Ethiopia Strategy Support Program and the Policy Studies Institute, Addis Ababa, Ethiopia.

Corinna Hawkes is director of the Centre for Food Policy at City, University of London, UK.

Derek Headey is a senior research fellow in IFPRI's Poverty, Health, and Nutrition Division, Washington, DC, USA.

Melissa Hidrobo is a senior research fellow in IFPRI's Poverty, Health, and Nutrition Division, Dakar, Senegal.

Kalle Hirvonen is a senior research fellow with IFPRI's Ethiopia Strategy Support Program, Addis Ababa, Ethiopia.

Neha Kumar is a senior research fellow with IFPRI's Poverty, Health, and Nutrition Division (PHND), Washington, DC, USA.

David Laborde is a senior research fellow in IFPRI's Markets, Trade, and Institutions Division, Washington, DC, USA.

Abla Abdel Latif is the executive director of the Egyptian Center for Economic Studies, Cairo, Egypt.

Will Martin is a senior research fellow in IFPRI's Markets, Trade, and Institutions Division, Washington, DC, USA.

Ruth Meinzen-Dick is a senior research fellow in IFPRI's Environment and Production Technology Division and co-leader of the CGIAR Research Program on Policies, Institutions, and Markets (PIM) Flagship Program on Governance of Natural Resources.

Bart Minten is a senior research fellow in IFPRI's Development Strategy and Governance Division, and former leader of the Ethiopia Strategy Support Program, Addis Ababa, Ethiopia.

Tia Palermo is an associate professor in the Department of Epidemiology and Environmental Health, University of Buffalo (State University of New York), Buffalo, NY, USA.

Amber Peterman is an associate research professor in the Department of Public Policy at the University of North Carolina at Chapel Hill, USA.

Mamata Pradhan is a research collaborator with IFPRI.

Agnes Quisumbing is a senior research fellow with IFPRI's Poverty, Health, and Nutrition Division, Washington, DC, USA.

Heather Rahimi is a communications and administrative associate of the Rural Education Action Program (REAP) in the Freeman Spogli Institute for International Studies at Stanford University, Stanford, CA, USA.

Delia Randolph is co-leader for Animal and Human Health, International Livestock Research Institute, Nairobi, Kenya.

Mariam Raouf is a research associate with IFPRI's Egypt Strategy Support Program, Cairo, Egypt.

Thomas Reardon is a professor in the Department of Agricultural, Food, and Resource Economics at Michigan State University, East Lansing, USA.

Danielle Resnick is a senior research fellow with IFPRI's Development Strategy and Governance Division, Washington, DC,USA, and leads IFPRI's Governance Theme.

Claudia Ringler is deputy director of IFPRI's Environment and Production Technology Division, Washington, DC, USA, and co-leader of the CGIAR Research Program on Water, Land and Ecosystems (WLE) Flagship Program on Variability, Risks and Competing Uses.

Sherman Robinson is an IFPRI research fellow emeritus, Washington, DC, USA.

Devesh Roy is a senior research fellow with the CGIAR Research Program on Agriculture for Nutrition and Health (A4NH), New Delhi, India.

Shalini Roy is a research fellow in IFPRI's Poverty, Health, and Nutrition Division, Washington, DC, USA.

Scott Rozelle is a senior fellow and the co-director of the Rural Education Action Program (REAP) in the Freeman Spogli Institute for International Studies at Stanford University, Stanford, CA, USA.

Marie Ruel is director of IFPRI's Poverty, Health, and Nutrition Division, Washington, DC, USA.

Claudia Sadoff is director general of the International Water Management Institute, Colombo, Sri Lanka.

Wei Si is a professor at the College of Economics and Management, China Agricultural University, Beijing, China.

Mark Smith is deputy director general of the International Water Management Institute, Colombo, Sri Lanka.

Gashaw Tadesse Abate is a research fellow with IFPRI's Markets, Trade, and Institutions Division, Addis Ababa, Ethiopia.

Seneshaw Tamru is a researcher with the International Growth Centre-Ethiopia, Addis Ababa, Ethiopia.

Agajie Tesfaye is a senior researcher with the Ethiopian Institute of Agricultural Research, Addis Ababa, Ethiopia.

James Thurlow is a senior research fellow with IFPRI's Development Strategy and Governance Division, Washington, DC, USA.

Maximo Torero is the chief economist and assistant director general for Economic and Social Development at the Food and Agriculture Organization of the United Nations, Rome, Italy.

Rob Vos is director of IFPRI's Markets, Trade, and Institutions Division, Washington, DC, USA.

Huan Wang is a research scholar with the Rural Education Action Program (REAP), in the Freeman Spogli Institute for International Studies at Stanford University, Stanford, CA, USA.

Michael Wang is a Mickey Leland International Hunger Fellow in IFPRI's Development Strategy and Governance Division, Washington, DC, USA.

Manfred Wiebelt is a senior research fellow and professor of Economics at the Kiel Institute for the World Economy, Kiel, Germany.

Yue Zhan is a research assistant with IFPRI's East and Central Asia Office, Beijing, China.

Yumei Zhang is a visiting research fellow with IFPRI and senior research fellow with the Institute of Agricultural Economics and Development of the Chinese Academy of Agricultural Sciences, Beijing, China.

Xiaobo Zhang is a senior research fellow with IFPRI's Development Strategy and Governance Division, Washington, DC, USA, a chair professor of Economics at Peking University, China, and a visiting fellow with the Center for Global Development and European Institute for Chinese Studies.

David Zilberman is a professor in the Agricultural and Resource Economics Department at the University of California, Berkeley, USA.

References

AGRA (Alliance for a Green Revolution in Africa). 2019. *Africa Agriculture Status Report 2019–The Hidden Middle: A Quiet Revolution in the Private Sector Driving Agricultural Transformation* (Issue 7). Nairobi.

Barrett, C., T. Reardon, J. Swinnen, and D. Zilberman. 2020. "Agrifood Value Chains Revolutions in Low and Middle Income Countries." Unpublished manuscript. (Appendix only: https://www.ifpri.org/appendix-estimates-importance-supply-chains).

Bellemare, M. F. 2014. "Rising Food Prices, Food Price Volatility, and Social Unrest." *American Journal of Agricultural Economics* 97 (1): 1-21.

Breisinger, C., A. Abdelatif, M. Raouf, and M. Wiebelt. 2020. "COVID-19 and the Egyptian Economy: Estimating the Impacts of Expected Reductions in Tourism, Suez Canal Revenues, and Remittances." Middle East and North Africa Regional Policy Note, IFPRI, Cairo.

Chen, Z., S. Fan, and Y. Zhan. 2020 (forthcoming). "COVID-19 and Food Security: Early Responses,

Impact, and Lessons from China," *China Agricultural Economic Review.*

Chou, E., and D. So. 2020. "Alibaba News Round Up: Helping Businesses Rebound from Virus." Alizila, June 5.

Coibion, O., Y. Gorodnichenko, L. Kueng, and J. Silvia. 2012, "Innocent Bystanders? Monetary Policy and Inequality in the U.S." NBER Working Paper 18170, National Bureau of Economic Research, Cambridge, MA.

Díaz-Bonilla, E. 2015. *Macroeconomics, Agriculture, and Food Security: A Guide to Policy Analysis in Developing Countries*. Washington, DC: IFPRI.

Díaz-Bonilla, E. 2016. "The Future Ain't What It Used to Be: Growth Models, Structural Change, and History." Markets, Trade and Institutions Division Working Paper, IFPRI, Washington, DC.

Díaz-Bonilla, E. 2018. "The Future Ain't What It Used to Be: World Growth and Global Governments." *Global Solutions Journal* 1 (3): 116-121.

Egger, D., J. Haushofer, E. Miguel, P. Niehaus, and M. Walker. 2019. "General Equilibrium Effects of Cash Transfers: Experimental Evidence from Kenya." NBER Working Paper 26660, National Bureau of Economic Research, Cambridge, MA.

Epstein, G. 2005. "Central Banks as Agents of Economic Development." Working Paper 104, Political Economy Research Institute, University of Massachusetts at Amherst.

Economic Times. "Flipkart Preparing to Start Hyperlocal Delivery Services." June 17, 2020.

Economic Times. "Amazon, Flipkart to Increase Warehouses." May 4, 2020.

Fei, S., and J. Ni. 2020. "Local Food Systems and COVID-19: A Look into China's Responses." City Regions Food System Programme, FAO, April 8.

Fettig, D. 2008. "The History of a Powerful Paragraph." Federal Reserve Bank of Minneapolis, June 1.

Financial Express Online. 2020. "Kisan Rath Mobile App: Centre Launches New App to Help Farmers During Coronavirus Lockdown." April 19.

Gilbert, M., G. Pullano, F. Pinotti, E. Valdano, C. Poletto, P.-Y. Boëlle, et al. 2020. "Preparedness and Vulnerability in African Countries Against Importations of COVID-19: A Modelling Study." *Lancet* 395 (10227): 871–877.

Handa, S., L. Natali, D. Seidenfeld, G.Tembo, and B. Davis on behalf of the Zambia Cash Transfer Evaluation Study Team. 2018. "Can Unconditional Cash Transfers Raise Long-Term Living Standards? Evidence from Zambia." *Journal of Development Economics* 133 (July): 42–65.

Harper, J. 2020. "Asia's Fishermen and Farmers Go Digital During Virus." BBC News, June 2.

Hidrobo, M., N. Kumar, T. Palermo, a. Peterman, and S. Roy. 2020. *Gender-Sensitive Social Protection: A Critical Component of the COVID-19 Response in Low- and Middle-Income Countries*. Washington, DC: IFPRI.

Hirvonen, K., G. T. Abate, and A. De Brauw. 2020. "Food and Nutrition Security in Addis Ababa, Ethiopia during COVID-19 Pandemic: May 2020 Report." Ethiopia Strategy Support Program Working Paper, IFPRI, Addis Ababa.

International Labour Organization. 2020. *ILO Monitor: COVID-19 and the World of Work*. Second edition Updated estimates and analysis. April 7.

Jinglin, Song. 2019. "Alibaba's Grocery Retail Chain Freshippo joins Ele.me to expand services: A Move to Go Beyond its Own App and Stores." KrASIA, September 12.

Kazeem, Y. 2020. "African e-Commerce Is Getting a Much Need Boost from Coronavirus Lockdowns." Quartz Africa, May 19.

Leesa-Nguansuk, S. 2020. "Covid-19: Thailand Eateries Banking on Delivery Apps, Social Media to Survive Lockdown." *The Star*, April 21.

Peterman, A., A. Potts, M. O'Donnell, K. Thompson, N. Shah, S. Oertelt-Prigione, and N. van Gelder. 2020. "Pandemics and Violence Against Women and Children." CGD Working Paper 528 (April), Center for Global Development, Washington, DC.

Reardon, T., R. Echeverria, J. Berdegué, B. Minten, S. Liverpool-Tasiea, D. Tschirley, and D. Zilberman. 2019. "Rapid Transformation of Food Systems in Developing Regions: Highlighting the Role of Agricultural Research and Innovations." *Agricultural Systems* 172 (June): 47-59.

Reardon, T., A. Mishra, C. S. R. Nuthalapati, M. Bellemare, and D. Zilberman. 2020. "COVID-19's Disruption of India's Transformed Food Supply Chains." *Economic and Political Weekly* 55 (18), May 2.

Reardon, T., and D. Zilberman. 2018. "Climate Smart Food Supply Chains in Developing Countries in an Era of Rapid Dual Change in Agrifood Systems and the Climate." In *Climate Smart Agriculture: Building Resilience to Climate Change*, edited by L. Lipper, N. McCarthy, D. Zilberman, S. Asfaw, and G. Branca, 335-351. Cham, Switzerland: Springer and FAO.

Rozelle, S., and J. Swinnen. 2004. "Success and Failure of Reform: Insights from the Transition of Agriculture." *Journal of Economic Literature* 42 (2): 404-456.

Swinnen, J., and M. Maertens. 2007. "Globalization, Privatization, and Vertical Coordination in Food Value Chains in Developing and Transition Countries." *Agricultural Economics* 37 (s1): 89-102.

Wolle, A., K. Hirvonen, A. De Brauw, K. Baye, G. T. Abate. 2020. "Household Food Consumption Patterns in Addis Ababa, Ethiopia." Ethiopia Strategy Support Program Working Paper, IFPRI, Addis Ababa.

Made in United States
North Haven, CT
05 November 2026

21766165R00082